YOUNG ADULTS
STO

**ACPL ITEM
DISCARDED**

Y 662.1 B75F   2254638
BRENNER, MARTHA.
FIREWORKS TONIGHT!

DO NOT REMOVE
CARDS FROM POCKET

ALLEN COUNTY PUBLIC LIBRARY

FORT WAYNE, INDIANA 46802

You may return this book to any agency, branch,
or bookmobile of the Allen County Public Library.

DEMCO

FIREWORKS TONIGHT!

FIREWORKS TONIGHT!

by
MARTHA BRENNER

Illustrated with prints and photographs

HASTINGS HOUSE PUBLISHERS, INC.

New York, NY 10016

Allen County Public Library
Ft. Wayne, Indiana

Copyright © 1983 by Martha Brenner

All rights reserved. No part of this publication may be reproduced, stored in a retrieval system, or transmitted, in any form or by any means, electronic, mechanical, photocopying, recording or otherwise, without the prior permission of the copyright owner or the publishers.

Library of Congress Cataloging in Publication Data

Brenner, Martha.
  Fireworks tonight!

  Bibliography: p.
  Includes index.
  Summary: Traces the history of fireworks and discusses some of the varieties available today, regulations, dangers, and safety guidelines.
  1. Fireworks—Juvenile literature. [1. Fireworks]
I. Title.
TP300.B65   1983   662'.1   83-8603
ISBN 0-8038-2400-9

Published simultaneously in Canada by
Saunders of Toronto Ltd., Markham, Ontario

Printed in the United States of America

Contents

2254638

In Appreciation	7
1. An American Tradition	9
2. Entertainment for Kings and Common People	21
3. Famous Fireworks Families	33
4. How Fireworks Work	51
5. Making Fireworks: Care and Craftsmanship	63
6. Getting Ready	77
7. Fireworks Tonight!	89
Fascinating Fireworks Facts	101
How to Watch a Fireworks Show Safely and How to Fire Home Fireworks Safely	106
Popular Home Fireworks	108
Illegal Fireworks	110
A Spectator's Guide to Fireworks	112
Selected Bibliography	116
Index	117

*For Saul Brenner,
who gave the gift of time.*

In Appreciation

Grateful acknowledgement is made to the following people who generously contributed their time and knowledge to this book:

Dr. John Conkling, American Pyrotechnics Association
Patrick Delville, Ruggieri—U.S.A.
Judy Donnelly and Julie Nord, Hastings House, Publishers, Inc.
Robert Funt
Paul Galvydys and Ann Pavlich, Consumer Product Safety Commission
Felix Grucci Jr., Fireworks by Grucci
Gary Partlow, Southern International Fireworks, Inc.
Nancy Rozzi, Tri-State Fireworks Company
Dr. Paul Saman, Carolina Translating and Interpreting Service
James Sorgi, American Fireworks Company
Bow Walker, Discovery Place, Science Museums of Charlotte
George Zambelli and Dr. George Zambelli Jr., Zambelli Internationale

As American as apple pie: A Fourth of July fireworks display glitters just above one of the capitol city's most notable landmarks, the Washington Monument. A Zambelli Internationale display.

An American Tradition

If you've ever waved a sparkler, lit a firecracker, or simply watched, amazed, as the dark sky suddenly burst alive with dazzling light, color and deafening sound, you are part of the long history of fireworks.

The fireworks you see today are more powerful and colorful than the simple firecrackers lit by the Chinese 1,300 years ago. But they are the same in many ways. Today, as long ago, fireworks are made by hand. Most contain the same basic ingredient used for centuries—gunpowder. Though fireworks are still dangerous and expensive, they continue to be a popular and exciting way of celebrating.

This book will reveal some of the mysteries of the fireworks you enjoy at big sky shows or on your own sidewalk. It will introduce *pyrotechnics*, the art of manufacturing and setting off fireworks, and the role of the fireworks technician, a professional known as a *pyrotechnist*. (Pyro means fire or heat and comes from the ancient Greek word for fire, *pur*.)

Triumph and Tragedy on the Fourth

It's hard to imagine a Fourth of July without fireworks. Even before the first Fourth of July, over two hundred years ago, revolutionary leader John Adams imagined the part "illuminations"—the name for fireworks in his time—would play in its observance. On July 3, 1776, he wrote to his wife that the next day would be "the most memorable in the history of America. I am apt to believe that it will be celebrated by succeeding generations as the great anniversary festival . . . with pomp and parades . . . bonfires and illuminations from one end of this continent to the other, from this time forward forevermore."

Lighting fireworks was one of the colonists' favorite ways to celebrate special holidays. The custom was brought over by the earliest settlers from England and Europe. In some colonies, there even were laws forbidding people from throwing "squibs" (little homemade firecrackers of gunpowder folded in paper) out their windows or at other people. Fireworks were a natural part of the first Fourth in 1777 and the grand celebration in 1789 when George Washington became the first president. Fireworks have played a role in American history ever since.

What John Adams couldn't have predicted, however, was the transformation of a national day of celebration into a national day of tragedy. By the late 1800s and early 1900s, the American public could easily buy and set off dangerously powerful firecrackers. So many terrible accidents happened that the Fourth was called "Death's Busy Day".

Yet, Americans ignored the risks and went right on buying bigger and louder fireworks. Sometimes manufacturers substituted deadly dynamite for gunpowder to make a louder blast. One favorite cracker was a foot and a half long! Dropped in an iron letter box, it could blow it to pieces.

Toy pistols with blank cartridges and toy cannons were popular—but no safer. The play "ammunition" could accidentally explode in a child's hand as he was loading the gun

Completed in the 1850s, this is one artist's view of an American family's Fourth of July.

or cannon. Earth and clay mixed with the gunpowder in the blanks got into the burn wounds. The dirt sometimes contained germs of an often fatal disease, tetanus (lockjaw).

On July 4, 1903, fireworks accidents occurred that injured almost 4,000 people and would kill 445. Of these victims, 406 died of tetanus. Doctors soon learned to treat fireworks patients with immunization shots against tetanus, but there was nothing they could do for the adults and children who were blinded or lost a finger or a hand in accidents. Sadly, innocent bystanders—sometimes little brothers and sisters who had nothing to do with the lighting of the fireworks—were the ones injured.

More than 4,000 people died in the United States from fireworks between 1900 and 1930—some while manufacturing fireworks, others while watching fireworks, and most while playing with them. That's more than the number of Ameri-

As part of the early 20th century campaign against the careless use of fireworks, this drawing appeared in a 1913 magazine with the ironic caption: "July fifth—And all's well!"

cans killed in the Revolutionary War. Celebrating our independence was costing more lives than winning it!

Was the excitement of fireworks worth such suffering? More and more people felt something must be done to protect the public safety. To dramatize the danger, The *Chicago Tribune* newspaper for years on July 5th and 6th published lists of fireworks injuries and deaths. The *Ladies Home Journal* magazine, early in the 1900s, urged parents to organize parades and sports events for children on the Fourth instead of buying them fireworks. By the 1930s many publications and organizations had joined in an anti-fireworks movement to convince state and city lawmakers to limit the manufacture, sale and display of fireworks.

Fireworks manufacturers were sternly warned by The National Society for the Prevention of Blindness, in 1937, to stop producing dangerous items before Americans "wipe out the entire (fireworks) industry and with it the barbarous practise of burning children alive, gouging their eyes out and blowing their hands off in the crazy notion that this is patriotism."

A 1930s cartoon.

Eventually, the lawmakers and the manufacturers responded. Some fireworks makers voluntarily stopped selling large firecrackers to the public. Others introduced safer fireworks for children that reduced the danger of fireworks exploding in their hands and faces. In the northeast, states banned even the smallest fireworks, such as sparklers and birthday party poppers.

Although no national agency or laws existed in 1938 to regulate fireworks safety, a model law was written for states to follow if they wanted. The law banned the sale and use of *all* fireworks to the public. Only licensed operators or special groups, like police or firefighters, could shoot fireworks. Today, 14 states have this strict law, but 36 states and Washington, D. C. allow the sale of some fireworks.

Protective Regulation

When you walk into a fireworks store and look at the dynamic names on the packages, the fireworks for sale seem strong enough to blast you to the moon. Fireworks called "Buzz Bombs," "Saturn Missiles" and "Thunder Buzzards"

A Cincinnati fireworks company advertised some of their wares in 1869 with these delicate drawings. Their names were: Ladies' Whims, right, Sparkling Caprice, opposite page, left, Turkish Cross, opposite page, right.

promise big-time excitement. There is even a brand called "T.N.T." after the compound used in heavy blasting.

Thrilling as these fireworks are to light in your backyard, they contain only a very small amount of explosive powder. Since 1976, the U. S. Consumer Product Safety Commission, with the help of the American Pyrotechnics Association (a group of manufacturers and distributors), has regulated the manufacture and labeling of fireworks for public sale. Only 50 milligrams of powder—just enough to cover your fingertip—are permitted in each firecracker. Fireworks that explode in the air, such as small rockets, can contain up to 130 milligrams. In addition to the explosive powder, manufacturers are permitted to add spark and color producing chemicals.

The Consumer Product Safety Commission also sets performance standards. Fuses on Class C fireworks, for example, must burn for at least three seconds (but no more than six seconds) to give the person lighting the firework time to get away. New fireworks on the market are tested to make sure they conform to the rules.

Federal government transportation laws divide all explosives into three categories: A, B and C. Legal fireworks for

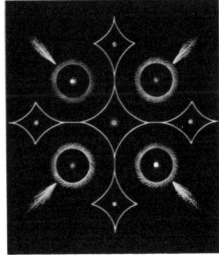

sale to the public are called *Class C* common fireworks, and are also known as home fireworks. State and local laws, which control the sale, storage and use of fireworks, also determine how old you have to be to buy them. The minimum age ranges from 14 to 18 years old.

The fireworks used by professionals are classified as *Class B* fireworks. These contain large amounts of explosive material and can only be set off at licensed public displays. Compared to a Class C firecracker, containing 50 milligrams of flash powder, a Class B noise-making shell—called a salute—might contain one ounce of flash powder or 28,400 milligrams! Military explosives, such as bombs and artillery shells, are *Class A* explosives. They are much more powerful than fireworks.

Mischief and Misuse

Despite these improvements in safety, fireworks injuries have not disappeared. A new high was reached during the Bicentennial year, 1976, when about 11,000 fireworks injuries were reported by hospital emergency rooms.

Fortunately, most fireworks injuries today are minor burns, not the explosive injuries of the past. Over half the accidents happen because adults and children use fireworks carelessly, or even mischievously, to frighten or hurt someone else.

Unfortunately, too, there are a few manufacturers who continue to illegally make and sell dangerously powerful fireworks which have been banned throughout the country since 1966. All of these contain large amounts of flash powder which explodes violently and loudly: the cherry bomb, a small red sphere one-inch wide; the M-80, used by the military to imitate real hand grenades (not to be confused with Class C smoke bombs or other legal fireworks with the same name); and silver salutes, a silver cylinder with a fuse coming out its side. Sold on the street or out of the back of vans, these fireworks contain about 25 times the explosive powder in Class C fireworks. They cause most of today's serious injuries.

Agents from the Federal Bureau of Alcohol, Tobacco and

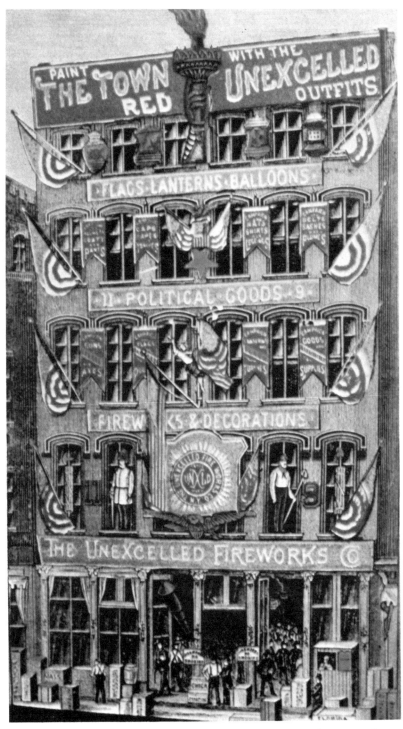

The 1888 catalog cover for "The Unexcelled Fireworks Company."

Firearms do the best they can to arrest these secret manufacturers, but they are hard to find, unless one of their hidden factories blows up. Bureau agents also must track down shipments of Class B fireworks that are lost or stolen on their way from a manufacturer or distributor to a buyer or the site of a show. A truckload of these high-powered fireworks in the wrong hands could cause a lot of damage.

How Safe Are Fireworks?

Safer chemicals, smaller amounts of explosives, better fuses and clearer warnings and instructions on labels have made home fireworks safer than ever in America's history. As a result, some states have changed earlier laws and now allow the sale of certain Class C fireworks.

But is safer safe enough? The National Society for the Prevention of Blindness, the Fire Marshalls Association of North America and other fire safety and health groups would like all home fireworks taken off the market.

On the other hand, Chinese-Americans and other ethnic groups have asked the federal government to give them special permission to use larger firecrackers for their religious celebrations. They want crackers loud enough to inspire "awe within the user." The Consumer Product Safety Commission turned down their request, but the Chinese developed a cracker wrapped in more paper that delivers a satisfying—and startling—level of sound with the legal amount of powder.

The fireworks industry strongly supports present government regulations. If fireworks are totally banned, legal manufacturers and distributors fear that illegal manufacturers would cash in on the opportunity and sell more dangerous fireworks, without any concern for safety.

Fireworks will never be harmless. As one government fireworks regulator puts it: "There is a low level of risk with everything." The problems of the past, however, have been greatly reduced. Although more Americans are using fire-

The great golden dragon parades through San Francisco's Chinatown during the Chinese New Year celebration. Fireworks are an essential part of this and other traditional Chinese-American holidays.

works, the percentage of the population being injured is far smaller than earlier in this century.

Since the Bicentennial, the popularity of fireworks is "booming." In 1981, American amateurs and professionals set off an estimated $130 million in fireworks. About 70 percent were home fireworks. Imports from the People's Republic of China are skyrocketing and stimulating a demand for new and unusual effects. The simple, old-fashioned firecracker now has competition from blazing spinning tops that jump and Jacks in the Box with sparkling surprises. The American tradition of fireworks—a tradition borrowed from other lands—looks like it has a strong future.

Entertainment for Kings and Common People

We don't really know who invented fireworks. Some historians think it was the Chinese. Others claim the ancient Hindus of India. We don't know who first produced gunpowder, the key ingredient in fireworks. Credit has been given to the Chinese, the Arabs, the Germans, the Greeks and the English. But no one is sure.

We do know that the Chinese were making primitive fireworks 2,000 years ago. They packed finely ground charcoal and sulfur into paper or bamboo tubes and attached a fuse. These early fireworks, however, produced only a flash of fire and a puff of smoke. We think the "bang" of the firecracker came later through trial and error.

The old story goes, a cook noticed that when saltpeter, a substitute for table salt, accidentally spilled in the fire under the cooking pot, the fire grew much brighter for a moment. What would happen if saltpeter, plentiful in northern China, were added to the charcoal and sulfur mixture already in use?

The result was a noisy explosion, the birth of the first firecracker, and the invention of gunpowder.

By the sixth century A. D., the firecracker was a familiar part of Chinese religious parades and festivals. Much later, in the 11th century, the Chinese got around to using gunpowder in weapons like bombs and rockets, but they never developed the gun. They were more interested in inventing new fireworks than weapons of war.

It's likely that the idea of gunpowder was brought to Europe in the 13th century by European soldiers returning home

Fireworks were also enjoyed centuries ago in India, as this antique miniature painting shows.

from the wars of the Crusades in the Middle East. In Jerusalem they had been impressed—and often defeated—by Moslem soldiers armed with guns and cannons. Europeans also were fascinated by stories about fireworks told by adventurers to the Far East, such as Marco Polo of Italy.

When Spanish and Portugese explorers visited China in the 16th century, they were amazed at how skillful and how extravagant the Chinese were with fireworks. (They also introduced the Chinese to guns and rifles.) Matteo Ricci, a Christian missionary traveling with the explorers, wrote in his diary that the Chinese were experts in "reproducing battles and in making rotating spheres of fire, fiery trees, fruit and the like, and they seem to have no regard for expense. . . . When I was in Nankin, I witnessed a pyrotechnic (fireworks) display for the celebration of the first month of the year (Chinese New Year) which is their great festival, and on this occasion I calculated that they consumed enough powder to carry on a sizable war for a number of years."

One of the first Europeans to write about gunpowder and firecrackers was Roger Bacon, a medieval English scholar. His 13th century book on science describes a child's toy as big as a person's thumb that exploded with the roar of thunder and flashed with the brilliance of lightning. He knew that saltpeter was the force behind this "horrible sound". Although guns weren't invented yet, Bacon realized the potential danger of gunpowder.

Cautiously, he wrote its formula in a secretive anagram—words that had to be unscrambled to be understood. His code was not broken for 650 years!

In the meanwhile, however, other Europeans were experimenting with different recipes for gunpowder. They were eager to develop it to its fullest force and apply it to new weapons. Eventually, they found a formula that created a powerful enough explosion to blast iron balls through a cannon toward a distant target. (Roger Bacon's formula wouldn't have worked as well.) Gunpowder gradually revolutionized warfare. The sword and armor of the knight and the arrow of the archer were no match for cannon and rifle fire. Castles

Roger Bacon (1214?–1294), whose gunpowder formula was written in a sort of code that was not deciphered for 650 years!

and walled cities crumbled under cannon attack. Only kings could afford to organize the large armies needed to wage a new kind of war. By the 1340s, many small factories called powderworks were grinding and mixing gunpowder to supply these new armies.

Fortunately for the history of fireworks, not all the gunpowder went into killing enemy soldiers. The military officers who became experts in handling gunpowder also contributed to the development of fireworks. When their armies won a battle, they were asked to provide a victory celebration display of rockets, cannon salutes and other fireworks. Sometimes a simple adjustment of a time fuse on a military rocket could convert it to a firework. As a weapon, the rocket would burst on the ground; for entertainment, it burst in the air at the highest point in its flight.

The job of putting on victory displays became so important by the 1500s, that Charles V, ruler of the Holy Roman Empire in Europe, kept fireworks specialists on his army payroll. These specialists, who were called firemasters, later transferred their talents to royal and religious celebrations. Some wrote manuals on warfare that included recipes, diagrams and instructions for presenting elaborate fireworks.

Fireworks, Italian-Style

The Italians were the first Europeans to develop fireworks into an art form. The 15th and 16th centuries, known as the Renaissance, were times of great artistic creativity and wealth in Italy. The Catholic Church and noblemen could afford to hire technicians to produce expensive spectacular fireworks displays for religious feast days and private celebrations. Some displays cost a fortune and burned for only a few minutes.

For beautiful shows in the sky, the Italians used aerial techniques that are still used today. The fireworks did not have color but produced sparks of gold and silver.

The Italians were best known, however, for their lavish productions on the ground. They would first create an elaborate wooden set carved and painted to look like a castle, palace or Greek temple. Then various fireworks were tied to it or set around it—torch-like flares, "fountains" that spewed forth streams of sparks and different sized pinwheels. Pinwheels are wooden wheels on long sticks with little rockets attached to them. When the rockets are lit the wheels spin and sparks shoot off, forming a pattern like a star or a sunburst. Pinwheels are still in use today.

With a variety of fireworks all blazing at once, an Italian "temple" was a spectacular sight. The Italian firemasters were in demand throughout Europe by princes and kings who wanted the best in fireworks for *their* crowning ceremonies, weddings, peace celebrations and religious festivals. Fire that could be made to perform exactly as wanted was a wonderful novelty.

Fireworks displays were often a kind of theater based on themes, mythical heroes and stories. Six hundred guests of France's King Louis XIV, for example, attended a three-day party of banquets, ballets, drama and fireworks all called "The

This 1810 Parisian fireworks display celebrated the Emperor Napoleon's marriage to Princess Marie Louise of Austria.

Pleasures of the Enchanted Isle". On the final night, they gathered in the gardens of the palace at Versailles. An Italian fireworks expert and his assistants ignited three fighting sea monsters and, as the monsters' flames faded, a splendid Palace of Enchantments blazed until it too vanished into the night.

In contrast to these Italian extravaganzas, fireworks shows in northern Europe, including Sweden, Denmark and the German states, were shot from the ground and burst high overhead, much like our modern displays. If a central "temple" was used, it was much simpler than the southern European structures. Instead, real buildings might serve as the center of a display, despite the danger of fire.

Flaming Ships and Dragons

Firemasters learned to magnify the effect of fireworks by setting them on little floats in fountains of water or at the edge of a reflecting lake or river. A ship or barge anchored off shore made a good stage for a fireworks spectacle.

On land or water, the skills of many workmen were needed to produce a fireworks show. Carpenters, wood carvers, painters, brickmasons, shell packers, metalworkers and Green Men, or Wild Men, were part of the firemaster's team. The Green Men walked at the head of royal processions waving fireclubs that showered the crowd with sparks and cleared the way. They dressed in grimy clothes, their faces were coated with soot and they wore hats that were covered with green leaves. All this was so that, in the dark, they would not be noticed as they ran about lighting the many fireworks at the display. But this strange get-up also made the Green Men very alarming to look at, and they often tried to seem even scarier by making wild noises as they paraded through the streets. They were as much a part of the show as the fireworks they lit at the end of the parade.

The favorite fireworks creation of the 16th, 17th and 18th centuries was the dragon. The dragons were a small marvel of art and engineering. They were built with rib cages of willow rods or whale bone. Their scaly skin was made of painted *pâpier-maché,* or isinglass, a transparent gelatin-like material. Small rockets were placed on each rib; larger rockets in the belly and tail. Rockets in the mouth made the monster look like it was belching forth flames. Fireworks artisans invented ways to make the dragon move on the ground or in the air along a wire, or through water on wheels.

Dragons were popular beasts at all kinds of royal occasions. The prince of France and his guests enjoyed a battle between two fire-breathing dragons on his fifth birthday in 1734. Elizabeth, daughter of King James I of England, was honored on the eve of her marriage in 1613 with a fireworks

A diagram from an artillery manual written in 1650 which offers "how-to" advice on building a set piece in the shape of a dragon.

drama starring a dragon, a lady in need of rescue, St. George (legendary dragon-slayer) and an enchanted castle.

Gradually, fireworks changed from the private pleasures of kings to public displays that drew huge crowds. Some rulers put on fabulous shows to increase their popularity. When Napoleon I crowned himself emperor, he ordered a great display on the Seine River in Paris. Barges held flaming set pieces (pictures in fireworks) of Napoleon's heroic crossing of the Alps Mountains on his way to military victory in Italy. The climax of the evening was the flight of giant balloons decorated with lanterns and a golden crown.

People from all classes of society came to British amusement parks to see fireworks in the late 1700s and early 1800s. They paid admission to thrill to the whirling wheels and flaring rockets attached to an ornate "temple". At some parks, however, the entertainment was more pitiful than beautiful. Small firecrackers were tied to bulls, dogs or bears who were set loose to run in panic through the laughing crowd.

More and Brighter Colors

Fireworks gained a variety of colors and a new brightness in the 19th century. Metal salts were added to fireworks powders to produce vivid reds, greens, yellows and blues. Many of the same salts are in use today. Extreme brilliance was achieved later in the century with other new ingredients—powdered or flaked aluminum and magnesium.

The production of strong colors was helped by changes in the basic explosive formula. By making use of certain new chemicals instead of saltpeter, fireworks makers were able to create a more violent explosion which made the metal salts burn more intensely and their colors appear brighter. (The "bang" was louder, too.) Manufacturing fireworks with these new chemicals was more dangerous. The slightest friction could trigger an explosion. Once a fatal blast occurred when a worker dipped a scoop coated with one chemical into a barrel of another.

A different trend in fireworks began to gain popularity in the last half of the 19th century: spectacular set pieces or "fire pictures". Although not entirely new, these huge portraits or scenes outlined in fireworks became bigger and unbelievably complex. They were a great success at a major fireworks competition in England in 1865 and thereafter became the new "temple" around which other fireworks, on the ground and in the air, were centered.

In London, people bought tickets to the grounds of a huge exhibition hall, the Crystal Palace, to see set piece "pictures" of real monarchs, contemporary heroes and historical events—instead of make-believe beasts and mythical heroes. As a special honor, a visiting king might be invited to pull an electrical switch and light his own picture in fireworks. The pic-

Some of the portraits in fireworks for which the Crystal Palace shows were well-known, including a favorite subject, Queen Victoria.

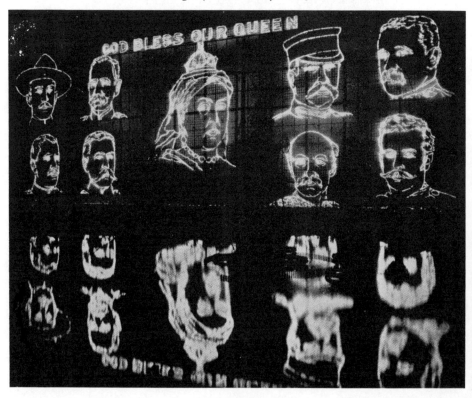

tures were made up of hundreds of tubes containing fireworks chemicals which burn with an especially bright, colorful light. These lances, as they are called, were mounted on narrow wooden slats and supported by large wooden frames hundreds of feet long, to form a single picture.

Audiences also liked humor and action. One favorite American set piece showed a donkey kicking a man whose head flew off and exploded. Another featured an elephant filling his trunk from a bucket and spraying flames. A fiery ten-foot walking hen who laid eggs delighted a crowd at Jones Beach, New York, in 1935. Each of the hen's three eggs hatched a chick. A fireworks assistant, protected by a tin shield, carried the hen while three helpers handled the chicks.

Also popular in the early 1900s was the *girandole* (pronounced jir-on-dol), a dazzling aerial firework. Technicians would raise it on a high pole from which rockets lifted it higher. People loved to watch it rise and spin like a flying saucer, dropping flames and changing colors. If a rocket failed or the wind shifted, however, spectators had to be careful. The girandole could come down right on top of them!

Fireworks makers helped their countries produce war materials in World War I and World War II since they were familiar with explosives. When peace came they turned to creating shells that could fly higher and deliver more spectacular effects. The new shells weren't cheap but they were less expensive than fancy set pieces that might be ruined in the rain before showtime. Aerial shells also didn't require a large open area for firing. That made them perfect for use in cities—large crowds could enjoy the display far from the firing site.

Sky Is the Stage

Gradually, the sky, rather than the ground, became the main stage for fireworks, just as it had been in 14th-century Germany and Sweden. Today's set pieces—the American flag, Liberty Bell or symbol of a show's sponsor—are only the supporting cast for the real stars of the show, the aerial displays.

Through the centuries, the sponsors of fireworks shows have changed too. Kings and rulers are no longer the most important patrons, although the 1981 wedding of Prince Charles of Great Britain to Lady Diana was royally celebrated with fireworks. Nowadays, in the United States, the sponsors who most often pay the bill for the shows we see are local department stores, banks, fairs, beer breweries, professional sports teams, college sports departments, civic clubs, shopping centers and newspapers. The amusement parks and exhibit halls of the past have been replaced by theme parks such as Walt Disney World which entertains visitors with fireworks year-round.

Fireworks are one of our most enduring customs and have spread all over the world. Perhaps they'll go even further, as space colonists set forth for distant planets. Someday we may even be setting off fireworks on Mars!

Fourth of July at Disney World.

Famous Fireworks Families

The fireworks industry is an unusual one. Most of the world's "major league" companies, who both make *and* shoot fireworks, are family firms. A fireworks company will often have three generations—grandparents, parents, children—working side by side and using family skills and recipes that have been handed down for a hundred years or more. Aunts, uncles, cousins, wives, husbands, and in-laws may be in on the act as well.

Children learn the family business from the cradle. First they travel to shows with their parents; later they master simple manufacturing tasks, pack shipments, and, by their late teens, they may be setting up shows. One fireworks company owner remembers putting protective caps on fuses when he was three years old.

A Special Heritage

What keeps fireworks a family business? Most fireworks

families are proud of their "fiery" heritage and take pleasure in passing it on to each new generation. Fireworks making may also have remained a family craft because there are few places to learn it—except for family firms. There are no technical schools that teach fireworks manufacturing, and no textbooks that explain more than basic fireworks chemistry. And, even for those born into the business, fireworks making takes years to learn.

A History of Competition

The fireworks business is highly competitive. The top manufacturers follow each other's work closely. One company will charter a plane after office hours to fly its top staff to watch a rival's show. As they fly home the same night, they analyze the show and "think how we could have done it better," says the owner.

Although the major firms produce shows of all sizes, their most important customers are a small group of fireworks users who spend $15,000 to $100,000 on shows for big cities and special events. If these entertainment organizers feel a particular company is experienced, creative, and reliable, they invite it to present a "script" or proposal. The proposal describes the different fireworks effects the fireworks company will design, and their prices. Sometimes cost is a deciding factor in being hired; sometimes not.

Family fireworks companies do cooperate with each other. The owners and their families meet annually as members of the American Pyrotechnics Association, a group that works on safety regulations and other matters. And, if one company should have a disastrous explosion at its plant, other firms will readily lend fireworks and other assistance to keep the stricken company going until it can recover.

But competition is almost a fireworks tradition in itself. In the 17th and 18th centuries, the cities and kings of Europe tried to outshine—and outspend—each other with impressive displays. Italian firemasters were especially famous for their skill and artistry, and many monarchs would invite them to

A 1732 Ruggieri fireworks display celebrating the birth of a Crown Prince to the King and Queen of France.

their countries to create new and greater masterpieces. The Italian visitors were not always welcomed by the local fireworks technicians, however. Jealousy and competition between the two groups could fan tempers to the boiling point.

At a peace treaty celebration in Paris in 1749, for instance, French and Italian fireworks experts argued over whose display would be lighted first. They couldn't agree. When they finally lighted both displays at the same time, a riot broke out and 40 people were killed.

Studying the generations of a fireworks family is like studying the history of fireworks. There are many fireworks families who have contributed to the art and technique of fireworks. These are some of the most active ones.

Ruggieri of France: High Fashion in Fireworks

The Ruggieri family is the oldest fireworks family in Europe and the Americas. They came from Bologna, Italy, but moved to France early in the 1700s. France's King Louis XV appointed the Ruggieri brothers "royal fireworks masters." Since then a Ruggieri spectacle—at the palace of Versailles or along the banks of the river Seine in Paris—has marked most of the big events of French history. In 1739, for instance, Ruggieri helped celebrate the marriage of Louis XV's daughter by staging a show all in silver and gold. A 170-piece orchestra provided the music.

Soon the name "Ruggieri" was a household word in France. It even took on a special meaning: if a woman was especially lively and flirtatious, people said she had a little "Ruggieri" (fireworks) in her.

The Ruggieri reputation spread to other countries. King George II of England hired Gaetano Ruggieri to create England's fireworks display commemorating a peace treaty. (The French show that broke out in a riot was a celebration of the same peace treaty.)

Grand "Temple" in Green Park

For nearly six months the Ruggieris worked with English master carpenters to construct the huge set piece "temple" they'd planned for the Green Park, London show. Decorated with statues, artificial flowers and painted pictures, the temple was as big as a five-story building. It was surrounded by 11,100 rockets, pinwheels, and other assorted fireworks. A famous composer, George Frederic Handel, wrote special music to be played during the show, including the firing of 100 brass cannons. (Today, this piece is known as the "Fireworks Suite".) The show *should* have been stupendous.

Instead, the grand performance turned out to be a grand disaster. Again, the native fireworks technicians and the visiting Italian-French specialists could not get along. They argued over how to light the show: by quick match—a covered fuse that burns extremely fast—or by pouring a trail of gunpowder. The show was delayed for hours. Finally, a trail of gunpowder was lighted causing an explosion and fire that de-

Modern Ruggieri fireworks, famous for their originality and artistry.

stroyed one whole wing of the temple. Afraid or impatient, the crowd began to drift away. When the display started, it was so late and so few spectators were still there that many of the fireworks were never lighted.

Afterward, British taxpayers complained about all the money wasted on the Green Park "show of the century". In the future, Ruggieri and other fireworks artisans produced less elaborate set pieces.

Vivid Colors and a Bull of Fire

In the 1800s, Ruggieri creativity turned in new directions. Claude-Fortuné Ruggieri was the first to write about the use of metal salts to produce color. He published a manual of color formulas in 1845. Today, a Ruggieri family member is active in the company, and many of its workers come from families who have been with the Ruggieris for generations. Establissement Ruggieri is still known for its vivid colors—especially green—and its slow-burning shells whose star-like fragments last longer in the sky. It can create novelty fireworks, too. One shell delivers 14 bursts of colored stars—seven on the way up, seven on the way down, aided by a tiny parachute that slows the shell's descent. When there is not much space for a fireworks show, Ruggieri provides a life-size "Bull of Fire". Six different fireworks effects sparkle and spin on the metal animal's head and back. It's something like a 20th century dragon.

The typical Ruggieri show is different in some ways from an American show. Dramatic, high-reaching ground fireworks alternate and blend with artistic combinations of aerial shells. "It's not just boom, boom, boom, sending shells in the air," explains a company spokesman. "Our concept is more like a play or ballet. And we think it is important to be close to the public. Smelling the powder and hearing the glittering adds a dimension that is lost in modern fireworks."

For America's Bicentennial in Washington, D. C., Ruggieri reproduced 18th century-style fireworks using the family's old formulas. One show honored Thomas Jefferson who admired the French company's work over 200 years ago on his trips to Paris. The show centered around a full-size canvas and wood model of Jefferson's home, Monticello. The display opened with red rockets and gold dust comets bursting in air while an orchestra played music of the period. Then Monticello was lighted, glowing in silvery lights. On either side, streams of fire plumed upward from Roman candles on the ground and flowers of green, gold, red and silver boomed high overhead.

For the Fourth of July, Ruggieri created a series of set piece scenes from America's history. Over 33 tons of fireworks were consumed. The price tag was $200,000!

The Brock Family: Gigantic Fireworks

England became a world leader in fireworks in the 1800s largely because of the Brock family. The founder of the Brock firm, Charles Thomas Brock, manufactured and exhibited fireworks for the British amusement parks in the 1700s. During the long rule of Queen Victoria, from 1837–1901, the Brocks won fame for the lavish productions they staged throughout the British Empire—and for the czars of Russia.

The Brock specialties were set pieces of historical scenes, well-known buildings and huge portraits of famous and honored people. One of their shows, "The Defeat of the Spanish Armada", featured a fleet of sinking ships; another, presented a realistic scene of firemen fighting a fire. For the most elaborate set pieces, they built wooden frames 80 feet tall and two football fields long on which they placed as many as 35,000 lances—little flares of varying size and color. Crowds of up to 80,000 people flocked to see these displays.

The Brocks created many fireworks portraits of Queen Victoria, but perhaps the most memorable—by mistake—was one at an anniversary celebration of her long rule. The portrait was part of a transformation set piece in which a colored floral design changed into glowing white portraits of the royal

SUPERIOR Fire works,

AT THE
ROYAL WILLIAM,
Bowling Green, and Tea Gardens,
IPSWICH,
ON TUESDAY AND WEDNESDAY EVENINGS, 7th and 8th of JULY; 1818,
BY
MR. BROCK,
ENGINEER,

Whose indefatigable study and attention to his profession, enables him to produce devices hitherto unknown to the Art, and with a richness and brilliancy surpassing all precedent.

PREVIOUS TO THE EXHIBITION IN THE GARDEN,
A GRAND DISPLAY OF VARIOUS KINDS OF
WATER FIRE WORKS,
ON THE RIVER.

Order of Firing.

FIRST DIVISION.

1. A Battery of Maroons, or imitative cannon
2. A Bengal light
3. Sky Rockets
4. A Saxon Wheel
5. Tourbillions
6. A regulating piece in three mutations, displaying a Vertical Wheel, a ... to ..., and a figure piece, of straw and brilliant fires
7. Line Rockets
8. Phenomenon Box and Mine
9. Sky Rockets
10. A Metamorphose with alternate changes, and a beautiful display of Chinese lattice work

SECOND DIVISION.

11. Sky Rockets
12. Horizontal Wheel with Roman Candles and Mine
13. Tourbillions
14. A regulating piece in two mutations, displaying a Vertical Wheel, and a superb display of chequer work, in straw and brilliant fires
15. Line Rockets
16. A regulating piece in three mutations, displaying a Vertical Wheel, changing to five Vertical Wheels, & a figure piece in straw & brilliant fires
17. Grand Battery of Roman Candles
18. A regulating piece in four mutations, displaying a Vertical Wheel, a brilliant Sun, five Cobourg Union Stars, and a figure piece of straw and brilliant fires
19. Sky Rockets
20. A regulating piece in three mutations, displaying a Vertical Wheel changing to a Pyramid of Wheels, and a superb shower of fire

ADMISSION, 1s. each.

Gardens open at half-past Seven, and commences at Nine o'clock precisely.

DECK, PRINTER, IPSWICH.

An 1818 handbill advertising fireworks displays by the English firm, Brock's Fireworks, Ltd.

Alan St. Hill Brock's memorable fireworks display, mounted for the 300th birthday celebration of Quebec in 1908.

family. As the blossoms faded and were replaced by Her Majesty's face, the wrong lances were burning. It looked as if one of her eyes was winking at the audience.

Recreating Niagara Falls

Arthur Brock came to New York in 1892 to prepare a monumental display to celebrate the anniversary of Columbus's discovery of America 400 years earlier. Brock used the Brooklyn Bridge—a man-made wonder of its day—as a set piece for recreating a natural wonder—the Niagara Falls—in fireworks. An audience of one million watched as rockets and shells were fired from either end of the bridge, while pinwheels and fountains were lit in the towers. Then, a Niagara of falling flares, a wall of fire, spilled down from the bridge's entire central span. Americans had seen fireworks replicas of the great falls on other occasions, but nothing matched this one.

Arthur Brock's son, Alan St. Hill Brock, outdid him with an even more gigantic show to celebrate the 300th birthday of Quebec in 1908. Along the St. Lawrence River huge portraits of famous explorers, generals and monarchs traced the French and British histories of the province in sparkling fire. At the climax, balloons with blazing fireworks were set aloft.

The Brock company is still thriving today.

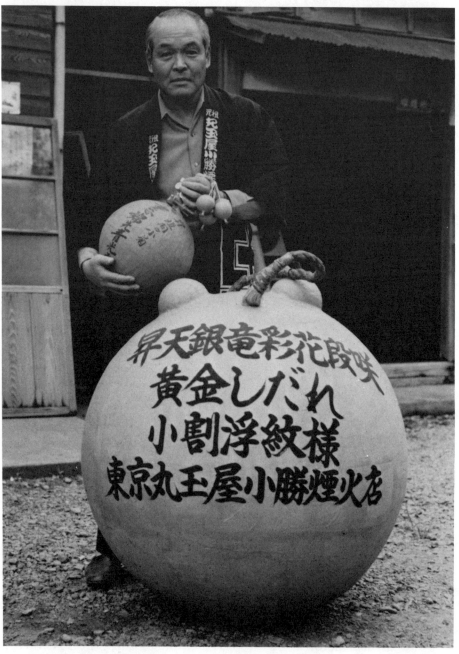

Kyosuke Ogatsu, Vice President of Ogatsu Fireworks, with a 36-inch Ogatsu shell.

The Ogatsu Family: Painting with Fireworks

The ancestors of the Ogatsu family created fireworks for Japanese war lords in the 18th century. Today, fireworks from Japan's Marutamaya Ogatsu Fireworks Company are known all over the world for their beauty.

Toshio Ogatsu, who designed many of the Macy's shows shot from barges in New York's Hudson River in the 1950's and 1960's, worked like an artist. First, he painted his ideas for aerial displays on canvas. He tried to capture his impressions of a flight of birds, the movement of fish or a waterfall. Using special manufacturing techniques, he and his company built shells that would reproduce these designs in fireworks.

Ogatsu carefully followed certain rituals whenever he presented a show. He came to the firing barge dressed in a ceremonial jacket. Around his head was a bandana covered with Japanese characters. Just before the first shell was lighted, Ogatsu would retreat to a corner of the barge and pray.

On a night when Ogatsu did not come on board the barges, a Japanese shell exploded a few feet above the deck and set off a whole barge full of shells. Two men from an American company firing the show were killed. The audience, unaware of the disaster, thought the fiery spectacle was part of the show. They applauded and cheered enthusiastically. "We certainly could have used Ogatsu's prayers that night," the president of the American company told George Plimpton, Fireworks Commissioner of New York.

Although Ogatsu knew little English, he communicated with his American technicians by sign language and sound effects. He was a fireworks show himself, they said, making pops, booms, hisses, and whistles to describe which shells he wanted placed in the mortars.

At one time, Ogatsu held the record for the world's largest firework. The shell was called the "Bouquet of Chrysanthemums Hanabi." ("Hanabi" is Japanese for fireworks. It means, literally, "flowers of fire.") It was 36 inches in diameter and weighed 200 pounds. It burst 3,000 feet over Tokyo Bay into a 2,000 foot wide display which changed colors many times. Ogatsu held his record until 1977 when the Grucci family of Bellport, New York, beat him out with Fatman II, a shell which was 40½ inches in diameter and weighed 720 pounds.

The Grucci Family: Record-Holders

Fireworks have been the business of the Grucci family of Bellport, New York for over 125 years. Felix Grucci, Sr. learned how to make and shoot fireworks from his uncle and has passed his skill on to his children, grandchildren and several other family members. His firm, New York Pyrotechnic Products Company (also called Fireworks by Grucci), is the only American fireworks company to have won the international fireworks competition in Monte Carlo, Monaco.

The Gruccis, who charge $1,000 to $1,500 per minute for a big show—minimum ten minutes—consider themselves entertainers. Their presentation on July 4th at the 1982 World's Fair in Knoxville, TN, demonstrated their flair for original showmanship. It also tested their "the show must go on" spirit. As the Gruccis were completing their preparations on July 3, a vicious hurricane-like storm swept through Knoxville damaging their sets. With a lot of hard work, however, they repaired everything by nightfall on the Fourth. The show went on! Besides ground flares and aerial shells, the main attractions were three glittering animated set pieces in which: the space shuttle "Columbia" blasted off; Little Orphan Annie waved her hand and Sandy the dog wagged his head; and seven children in costumes of many countries danced to "It's a Small World". Fireworks technicians behind the scaffolding made the "dancers" move and jump.

In addition to major shows like the World's Fair display, the Gruccis have supplied fireworks for New York Mets baseball games, Boston Pops concerts, motion pictures, television specials and commercials for Pepsi-Cola, McDonald's restaurants, and Macy's. And, they created twelve evenings

The Gruccis most recent pyrotechnic marvel was the Brooklyn Bridge Centennial display in 1983, pictured above.

of fireworks displays for the 1980 Winter Olympics at Lake Placid. They also like to develop new special effects such as their green and white strobe shell which goes on and off in the sky like a strobe light.

In one way or other, the Gruccis are known for creating a big bang. For one customer, they created a "bomb" 24 inches in diameter that cost $1,000. The monster created a brilliant cascade that covered 2,000 feet of sky and lasted for four seconds.

The Gruccis are famous, too, for their bombardment finales—the thunderous, non-stop noise at the end of a show. At one 25-minute show on July Fourth in Washington, D. C., the Gruccis shot one-third of their 2,300 shells in the last two and a half minutes.

The Sorgis: Staging an International Event

James Sorgi, head of the American Fireworks Company in Hudson, Ohio, is the fifth generation of his family in the fireworks business. His son, John, is the sixth. The Sorgis still make fireworks based on old recipes handed down from James Sorgi's grandfather to his father, Vincenzo "Jimmy" Sorgi, who came to American in 1902.

Before Vincenzo Sorgi opened his first fireworks company in Hudson in 1910, fireworks were just his hobby. The townspeople often heard him down by the railroad tracks setting off explosives; they nicknamed him "Jimmy the Bomb". His experiments with fireworks soon led to rocket building. When he announced that he wanted to build a rocket powerful enough to fly to the moon, daredevils from all over the world wrote to ask him for a chance to be the first man in space.

Like modern space ships, "Jimmy the Bomb's" rockets used a series of timed explosions to provide continuous power. One reached a height of 1,000 feet—higher than the Eiffel Tower—before crashing. Another went up, out of sight, and was never found. He wanted to send his dog for a ride in his next rocket and designed an explosive device to open the parachute that would float his pet safely to earth. But the town's dog owners protested so loudly, he gave up the idea. Many years later Mr. Sorgi welcomed the successes of the NASA space program. "I started something," he said. "I'm glad they are finishing it."

"Jimmy the Bomb's" son James is more down to earth about fireworks. Every year, he and his assistants truck about eight tons of fireworks (over 2,500 shells) to Detroit for the International Freedom Festival celebrating both Dominion Day in Canada and the Fourth of July in the United States. The display is set on two barges in the river and is controlled from an electric switchboard on a nearby tugboat. The show draws more than a million spectators.

James Sorgi, President of the American Fireworks Company, and son of Vincenzo "Jimmy the Bomb" Sorgi.

In 1964, the Democratic Party gave Sorgi an unusual assignment. It was almost a big flop. They wanted to wish President Lyndon B. Johnson a happy birthday with fireworks. Sorgi came up with a grand show: a 900-square-foot set piece portrait of Johnson and a 600-square-foot birthday cake. He set it all up on the beach outside the Atlantic City hall where the President was coming to make a speech at the Democratic National Convention. Then he waited for the president to arrive; that would be his cue to light the "cake".

He waited. And waited. As Sorgi counted the minutes, he noticed—with alarm—that the tide was rising higher on the shore. If he waited much longer, the waves would be lapping around the fireworks, and President Johnson's big birthday surprise would just fizzle and die.

Sorgi sent an urgent message to the President to come to the beach right away. "Luck was with us," he recalls. The President beat the tide and received his brightest birthday cake ever.

The Zambellis: Skills of Three Brothers

Three brothers—Louis, George and Joseph Zambelli—operate one of the oldest and largest of American firms that both manufacture and display fireworks, Zambelli Internationale. The company was started by their father Antonio at the turn of the century near Naples, Italy. In 1921, he moved his business and his family to New Castle, Pennsylvania. New Castle was once the home of 25 percent of America's fireworks industry. Over the years, restrictive laws and competition from imported fireworks have closed the doors of many of its plants. But not the Zambellis'.

Fireworks are made year-round at two Zambelli plants under the careful supervision of Joseph Zambelli, superintendent of manufacturing, and Louis Zambelli, vice president. At times, cousins who make fireworks in Italy come over to work with them. President George Zambelli manages the business side of fireworks, traveling around the country to

The Rose Bowl is one of the many major American public events for which Zambelli Internationale provides its famous, dazzling fireworks.

sell shows and design them. Two of his daughters work in the office and his son, George Jr., an eye doctor, keeps in close touch with the business, often even pitching in full-time for a few weeks during the busy summer season.

"Fireworks are alive with culture," George Zambelli believes and, like other forms of art, can be adapted to serve any purpose or idea. The Zambellis, for example, have a ten-foot-high Dancing Model set piece with movable arms and legs. They can "dress" the model with fireworks to look like a man or a woman. Sometimes the Dancing Model wears the native costume of a group celebrating its nationality day.

The Zambellis have added the excitement of fireworks to press conferences for designer clothes, the opening of new buildings, sports events and concerts for such popular stars as Elton John. They created a giant 80 x 100-foot Boy Scout emblem in fireworks for a National Boy Scout Jamboree. At the 1982 Orange Bowl half-time show they combined fireworks with a full symphony orchestra and computer-controlled light boxes—boxes on the field mounted with lights that can be programmed to change patterns. At one point, four light boxes shaped like tanks rolled in from the corners of the football field and shot fireworks at each other. The crowd loved it.

Behind the action on the field and in the sky were Zambelli technicians who had synchronized the firing of the shells electronically to a musical tape of the orchestra's concert. They didn't take any chances with their part of the show. A back-up crew stood ready to fire the shells by hand if the electronic ignition system had any failures. "We only had eight minutes. Everything had to work on cue," recalls George Zambelli.

To him, "the sky is the world's best backdrop." In the future, the Zambellis would like to help fine artists "paint" on this backdrop with fireworks custom-made to their designs.

How Fireworks Work

Have you ever played with a Jack-in-the-box? To make it work, you had to push Jack's puppet head down into his box. His head was attached to a wire spring which compressed into a tight coil as you pushed the head down. You had to shut the lid fast to prevent Jack from springing up.

Like a Jack-in-the-box, an unlighted firecracker is full of energy ready to spring free. Unlock the lid of Jack's box and out he pops. Put a match to a firecracker fuse and—bang—it bursts its paper tube. The surprise of a firecracker and a Jack-in-the-box are so similar, in fact, that a small firework named "Jack-in-the-Box" has been around since the 1600s. It starts off quietly with a spray of sparks, then produces a second explosion that blows dozens of tiny crackers or fissing poppers with tails into the air.

In a firecracker, burning releases *chemical energy* loosely "locked" in an explosive powder, usually black gunpowder.

A flame or an electric spark starts a change happening in the powder's ingredients: saltpeter (potassium nitrate), charcoal and sulfur. The bonds that hold the original elements together are broken. As atoms from one chemical combine with atoms from another chemical to form new molecules, heat and gas are produced.

Now, if gunpowder is poured out on the ground and lighted, there will be no explosion. No bang. Instead, the powder will burn with a flame that travels faster than your eye can follow. (WARNING: gunpowder is classified as a low explosive but it is still dangerously hot when it burns. Never play with it.)

When gunpowder burns in open air, the heat and gas generated can escape quickly. Confined in a heavy paper firecracker tube, however, the powder burns more rapidly, making the temperature zoom. The chemical change or reaction speeds up. Suddenly, gas is produced and rapidly expands to fill the container. Gas molecules fly around wildly and push out at the insides of the tube until the tube bursts from the pressure. There is an explosion as the trapped gas and heat rush out, setting the air in motion. You hear a big "bang".

A similar explosion takes place when you make popcorn. As you heat the popcorn kernels, moisture inside each kernel turns to steam—a gas. At first, the kernel's shell-like outer hull prevents the steam from escaping. But pressure builds until the hull bursts. Pop! The kernel explodes into puffed corn.

Gunpowder, The Basic Ingredient

The formula for gunpowder has changed little since the 16th century when armies first relied on it for their guns and cannons. Early in the 17th century, an Englishman named John Bate wrote a book on making fireworks. He explained the purpose of each ingredient in gunpowder this way: "The Saltpeter is the Soule, the Sulphur the Life, and the Coales the Body of it."

THE
SECOND BOOKE

Teaching most plainly, and withall most exactly, the composing of all manner of Fire-works for Tryumph and Recreation.

By IOHN BATE.

LONDON,
Printed by *Thomas Harper* for *Ralph Mab*.
1635

The title page of John Bate's book, published in 1635, features a "Green Man."

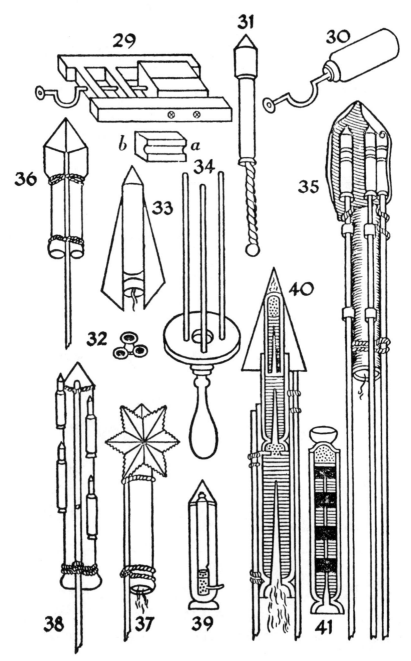

From a book published in 1747, this illustration shows several designs for 18th century fireworks.

In other words, the saltpeter provides the oxygen—the "soul"—that mysterious, invisible thing needed for burning to take place. The sulfur first catches the fire or "life" and passes the heat quickly from particle to particle, without being burned up itself. The coal is the fuel or "body" that burns.

Chemists and fireworks manufacturers can control the rate at which gunpowder burns by: (1) confining it to a container or letting it burn in the open air (important factors in creating fast and slow burning fuses); (2) changing the amount of each ingredient in the formula; (3) changing the size of the powder particles. Fine grains of black powder will burn fast and hot. Coarse grains will burn more slowly. Less saltpeter and more charcoal slows the burning and creates glowing sparks.

For example, two kinds of gunpowder are used to make small Roman candles, a firework available to the public in many states. Roman candles are long tubes that you stand upright on the ground. When lit, they shoot out high-flying bursts of colored fire and low, fizzing showers of sparks. Several small explosions of regular gunpowder propel the colored bursts into the air while slower burning candle comp, a gunpowder composition containing extra charcoal, makes the glowing sparks.

In a large fireworks shell shot by professionals, several gunpowder explosions are needed to make the different effects happen. Each quantity of powder, or other explosive, to be set off at one time is called a *charge*. Gunpowder serves as:

- a *lifting charge* to propel the firework shell high in the sky;
- a *bursting charge* to ignite and scatter chemical "decorations"—sparks or colored stars—packed inside the shell;
- a *sound charge* to create a small or big bang.

For some purposes, modern fireworks makers modify the original gunpowder formula by substituting other chemicals. Or they use entirely different explosives. For example, perchlorate mixtures are used in maroons—fireworks with a sharp,

An aerial display shell exploding.

very loud bang. Traditionally, maroons are the first shells fired in a show because their awesome noise immediately captures the audience's attention. Their name comes from the French word for large chestnuts, *marrons*. (Chestnuts sometimes explode while being roasted.) Chlorates produce a more violent explosion than black gunpowder. This results in a bigger crack of sound.

Blasting Out the Cargo

During a big fireworks show, streams of sparks and sprays of colored stars fly from "bombs" called *aerial display shells*. Aerial means high in the air—some shells reach 1,000 feet. Display means these fireworks are visible—you hear *and* see them.

Basically, an aerial display shell is a heavy paper cylinder or ball filled with several pounds or more of explosives and a "cargo" of burnable materials that produce special effects. The

The 16-inch chrysanthemum shell shown here is one type of large aerial display shell. Aerial display shells can also be cylindrical and range from 3 to 12 inches in diameter. Shells that are any larger than this 16-inch one are rare.

cargo is often made up of dull-looking cubes of chemicals and metallic powders, called *stars*. Looking at them you'd never guess that, heated to the right temperatures, they turn into brilliant flashes of light or cascading bouquets of rich color.

A shell's stars are contained in separate cardboard compartments within the shell. Each inner container has its own fuse and a *bursting charge* which lights and throws out the stars. In order to spread these decorations over a wide area of the sky, the container must burst open with tremendous force. The more the container can resist the explosion of the bursting charge, "bottling up" its force, the bigger the display will be.

"You are doing complete opposites when you make a shell," explains fireworks manufacturer George Zambelli. "You create an explosion and you contain it by creating resistance to it." Resistance comes from the star container's heavy wrapping, which is strong enough to momentarily trap the gas and heat from the bursting charge. When the gas finally rushes out—breaking the container—the burning stars are hurled into space with full force.

Usually the bursting charge is a small quantity of fine black gunpowder, but at least one fireworks maker says when he wants to light stars more brightly and create a faster, far-reaching spread, he inserts a bag of a highly explosive chemical mixture called flash powder (a perchlorate mixture) into the star compartment. Fireworks makers can further control the "look" of an explosion by adding more flash powder or adding more restriction.

Miraculous Multi-Break Shells

Americans are known for their aerial display shells built like multi-stage rockets. In a single shell, there may be three to five inner containers or *breaks*. Each break must explode separately at just the right moment in the shell's flight to make a beautiful display.

The flight of the average aerial shell begins when it is loaded into a launching tube called a mortar, which is about twice as long as the shell. The mortar is similar to mortars used to shoot military bomb shells, except that it is aimed straight up at the sky. A long fuse attached to the top of the shell hangs out the open end of the mortar. Lighting this primary fuse also lights a fast-burning side fuse that carries fire to the lifting charge at the bottom of the shell.

As the lifting charge explodes, hot air rushes downward out of the bottom of the shell. Thousands of hot air molecules bounce off the insides and end of the mortar and push the shell in the opposite direction—up! With a whoosh, the shell takes off. (A basic law of science is at work: for every action, there is an equal and opposite reaction.)

The primary fuse continues to burn as the shell shoots upward. When the shell reaches more than halfway to its peak height, the primary fuse has burned low enough to ignite the time fuse sticking into the first break. Now a series of timed explosions starts.

Fire hits a bursting charge of gunpowder in the first break. Bang! Color stars burn and scatter. The explosion sets off the time fuse leading to the next break. In a few seconds, the second break explodes, sending more stars into the sky. As the rocket heads downward, the third section blows, then the fourth. After a second's pause, you hear the "last word"—the thunder of the *sound charge*.

Timing is important because a break that explodes too soon or too late may not throw its stars high enough in the sky for spectators to see. Also, without slight pauses between explosions, the orderly beauty of the show would turn into noisy confusion. All the explosions of the shells are controlled by fuses of an exact length, containing an exact amount and fineness of powder.

Planning and construction of timed fuses is an important part of the artistry of fireworks. It is an art acquired by patient testing and re-testing.

Side Effects: Sound and Shock Waves

That "last word" of the display shell—the *sound charge* —is a final white flash and a tremendous boom. What's happening to produce a sound you can hear for miles?

When a sound charge explodes, heat is released, setting the air around it into motion. The heated air molecules move very fast, bumping other cooler and slower moving air molecules and speeding up their movement. They, in turn, bump other slower moving air molecules.

Almost instantly, compression waves of moving air spread out from the explosion, much like the ripples in a pond where you've just thrown in a stone.

When these waves move through the air near you, they vibrate your ear drum vigorously. Your hearing nerves transmit a loud "bang" to your brain. The bigger and hotter the explosion, the more air is moved and the stronger the sound waves.

Sometimes a gigantic explosion will produce larger and more powerful compression waves called shock waves. Shock waves can be destructive. Fatman II created a shock wave that broke 60 windows in the town of Titusville, Florida, about 15 miles from the launching. (You can understand why fireworks display companies carry one million dollars worth of insurance for each show. They may have to replace a lot of windows.)

The creators of Fatman II—the Grucci family of New York Pyrotechnical Products Company—could actually *feel* the shock wave. Perhaps Fatman II was too fat because all the sections of the shell exploded at once when it was barely out of its mortar. Watching from a safe distance upriver from the island launch site, the Gruccis saw a low ball of orange fire. Then they felt a soft wind cross their faces as the shock waves moved inland.

Far more destructive shock waves resulted from a terrible fireworks accident in 1902 in New York. William Randolph

A newspaper illustration showing one of many fireworks disasters: the explosion, in 1858, of a fireworks factory—Madame Coton's—in London.

Hearst, a wealthy newspaperman who had just been elected to Congress, held a victory celebration in Madison Square Park. Fireworks were to be part of the evening's celebration. Thousands of shells were stacked in piles, ready for the show. Mortars were set out on the ground. At one firing site at the edge of the park, people crowded around, as close as four feet from the mortars and shells.

As the fireworks began, one shell exploded too soon, knocking over its launching mortar and other mortars, too. Fire spread, and suddenly shells from the mortars shot directly into the crowd and set off explosions in piles of shells nearby.

The heat and shock waves took a heavy toll. Eighteen people were killed and nearly 100 wounded. Many windows were smashed in the city blocks surrounding Madison Square. The effects of this nightmare were felt for years, as injured victims and families of the dead sued Hearst. It was never clear who was to blame for the accident. At least one safety measure was not taken. The mortars probably should have been placed in pits in the ground.

The Hearst tragedy shows the damage many large fireworks can do. But even the explosion of one firecracker, like the now illegal cherry bomb, can easily destroy a small solid object nearby.

The hot air and gas released during a firecracker's explosion send hundreds of thousands of rapidly moving air molecules ramming against an object. They make it terribly hot. The force of these ramming molecules is stronger than the force holding the molecules of the object together. The object is torn apart, just as two magnets can be pulled apart by a force stronger than their magnetic attraction.

Imagine what such a force could do to you.

Fireworks professionals never forget how dangerous fireworks can be. Whether they are making them, transporting them or firing them, they take great care that a firework explodes at the right time in the right place.

Making Fireworks: Care and Craftsmanship

Unless you work at a fireworks manufacturing plant, you'll probably never visit one. Companies don't want visitors anywhere near their explosives and most plants don't look too friendly. They are surrounded by high chain-link fences topped with rows of barbed wire. Locked gates and danger signs warn passersby to stay away. There probably won't even be a sign identifying the company.

Inside the fenced area is a scattered group of small one-room buildings, sheds, larger storage buildings, trailer trucks and a distribution center with a high loading dock that trucks back up to. The buildings are spread far apart from each other so if an explosion or fire occurs in one the whole plant won't blow up.

In many states, laws regulate where and how fireworks plants are built to make them safer for the workers and for the people who live nearby. One law requires that fireworks storage buildings be fire-resistant, weather-resistant and theft-

Signs like these are a common sight where fireworks are manufactured.

resistant. Steel doors and special tamper-proof locks are common protective measures.

Fireworks manufacturing is an industry of small businesses in which workers do nearly everything by hand—or foot—or with their teeth (tying string)! Little machinery, if any, is used because of the danger of metal machinery producing heat or a spark that could ignite the explosive materials. Also, oil from a machine could combine with certain chemicals in fireworks to form an explosive gas.

The manufacturers of home fireworks, who produce half the fireworks made in America, stick to hand methods for another reason. They are reluctant to buy machinery as long as there is still a slight chance that federal laws could change and stop the sale of fireworks to the public. If this happened, they would be out of business.

Even working by hand, with the simplest of tools, workers must take special safety measures against the constant danger of explosions. No smoking is allowed wherever fireworks are handled, of course. All light bulbs are enclosed in protective wire covers in case one "pops". Fireworks—parts or completed shells—are never *pushed* across a table. They are always *picked up* to avoid creating one of the fireworks maker's deadliest enemies, static electricity. (This is the same electricity that sends sparks flying from your hand when you touch a metal door knob after walking on a thick rug.) In a fireworks workshop, the slightest friction or rubbing of one object against another could build up an electrical charge with its hazardous sparks. Fireworks craftsmen avoid wearing nylon or rayon clothing because it produces static. Instead, you'll see a lot of faded cotton blue jeans and T-shirts being worn at a fireworks plant.

Static electricity was the suspect in an accident at the Zambelli Internationale plant many years ago. The president's brother-in-law was working alone in a small storage shed. There were no witnesses, but family members think he may have scraped a ladder across the floor, creating a spark. He was killed when the fireworks in the shed exploded. Today, the Zambellis and other fireworks manufacturers pay hundreds of thousands of dollars each year for life and accident insurance to protect their workers and their workers' families.

European vs. Asian Style Shells

Only a small number of firms make display fireworks and of these only a few family-run businesses both make and shoot fireworks for major shows. The aerial fireworks they make are European style: they stack tubular containers or breaks on top of each other and wrap them in heavy brown paper. The breaks—loaded with stars—explode one after the other in brilliant sprays of color and thunder claps of sound.

To put more variety in a fireworks program, however, American display companies also buy shells from countries in Asia, Europe, South America, and from Australia. The

People's Republic of China, legendary birthplace of fireworks, is America's biggest supplier by far of foreign-made fireworks.

Asian or Oriental fireworks are shaped like a ball, rather than a tube. That is the secret of the perfectly round patterns they form in the sky. Deep inside the ball is a bursting charge surrounded by many carefully arranged layers of stars. The shells explode from the inside out creating a spreading flower shape, but little noise. Often these shells have the names of flowers: chrysanthemum, peony with pistils, wisteria and narcissus.

American makers have the skill to duplicate these Oriental beauties, but building such complex shells is very slow and costly work. In Asian countries, like Taiwan, China and Japan, where wages are much lower than in the United States, manufacturers can pay for long hours of labor and still sell their products at a profit.

The multi-break aerial display shells that are the pride of American manufacturers also require long hours of steady, careful and repetitious work. Making their own shells, however, gives fireworks companies more control over the final results in the sky, greater safety and more independence, in case their foreign supply is suddenly cut off or becomes too expensive.

A Visit to the Zambelli Factory

The main plant of Zambelli Internationale, which manufactures display fireworks only, is on a peaceful 200-acre hillside in western Pennsylvania. Work goes on all year, but the best seasons for making fireworks are late spring, summer and early fall. The good weather permits the workers to dry fireworks parts outside. Warm days also allow employees in the dustiest jobs, like mixing explosive powder, to work in open sheds with plenty of fresh air.

Star Mixing Room: Bare Hands for Tools

On a sunny, mild October day, two Zambelli technicians are mixing powder to make red stars that will go into multi-break shells. These are the shells that fling out fiery, colored stars in a series of timed explosions. The technicians mix one color a day.

The young assistant measures out scoops of chemicals from barrels onto a scale. One ingredient—the "oxidizer"—makes the stars burn fast and hot. Another—the coloring agent—

Mixing ingredients by hand to make red star compound.

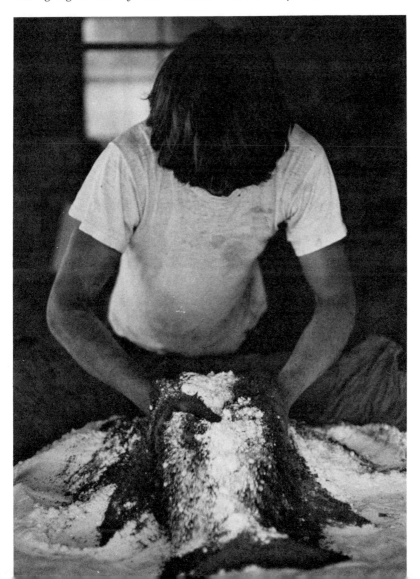

creates the brilliant red (or green, or gold) you see when they explode. A third chemical binds all the ingredients together.

The powders are sifted twice through a screen of brass, which won't create sparks. The ingredients for each batch are piled on a large sheet of paper and gently but thoroughly mixed by hand. There are no electric mixers, not even a wooden spoon. Friction must be kept to a minimum.

Other manufacturers use similar ingredients but their colors may look different in the sky. "The ingredients are only ten percent of the task of reproducing an effect," explains John Conkling, a college chemistry professor who serves as a technical representative for the American Pyrotechnics Association. How fine the chemical powders are, what binder is used and the order in which the ingredients are blended and mixed can change the end results. "There's a lot of art involved," says Dr. Conkling.

The Cutting Room: Dicing Stars

When a barrel-full of red star compound has been mixed by Zambelli technicians, it is taken to the next building, the Cutting Room.

The Cutting Room is a grimy kitchen-like place with a sink, work counters, mallets, molds and knives. A layer of sawdust covers the floor, a Zambelli safety measure to reduce the friction of shoe soles scraping against the floor. The three Cutting Room "cooks" are working with a recipe that could get very hot, indeed.

Their first step in forming stars is to add water to the barrel of red compound. Again, the dampened chemicals are gently mixed by hand until they form a clay-colored mud or dough. The dough, 35 pounds at a time, is scooped into a paper-lined wooden mold and pounded down into the shape of a large loaf of bread. The technician can safely pound the dough with his wooden mallet because the chemical mixture is wet and is not likely to explode from a hard shock or spark.

Below left: *A fireworks technician slices redstar compound "dough." The slices shown here are next cut up even smaller—into cubes, or "stars,"* top. *Then the stars, which will someday produce colorful bursts of light in the sky, are placed in the sun to dry,* below right.

The dough is unmolded on a work counter lined with heavy cardboard that has been sprinkled with fine black gunpowder. One man slices the dough; a second man dices the slices into small cubes as fast as a master chef. The cubes—or stars—become covered with gunpowder which will act as a primer to help them start burning. They are placed on paper-covered screens and set outside to dry.

At the end of every work day, the heavy cardboard that covers the work counters and all the paper used in mixing and molding the explosive compounds is thrown away. New paper and cardboard will be laid out the next day. In this way, all working surfaces are kept clean, preventing the contact of chemicals that may react suddenly in combination and cause an explosion. Explosives can behave unpredictably, even in the hands of people who know how to work with them.

Cutting Room workers like to say they owe their safety to "no smoking and a lot of prayers".

Spiking Room: Making the Breaks

After the star cubes dry, they are trucked to a small plant nearby—the old Zambelli plant—to be put into breaks, the small containers that make up a fireworks shell. A team of three men led by Joseph Zambelli, 74, superintendent of manufacturing, is preparing the parts for a shell that will produce two blasts of color stars and one ear-ringing "report" or boom. These three craftsmen have more than 150 years of fireworks making experience among them.

One of the craftsmen (who has made fireworks in Italy and England and says working at Zambelli is not much different) is filling the cardboard break containers with dried stars. He holds a hollow tube in the center of the container and pours handfuls of stars around it. It may take as many as 900 stars, or one pound, to fill a large break. Filling the break is done very gently and by hand. The craftsman is careful not to drop any stars or pour them too quickly. A sharp blow could ignite them and he doesn't want to see these stars twinkling prematurely.

After the stars are in place, he loads the center tube with gunpowder, then lifts the tube out. The powder is the bursting charge that will light and spread the stars. He completes his job by putting a paper cap on the filled container and passing it along to the next man for *spiking*.

The spiking man sits behind a huge spool of heavy string. He ties string from the spool to the container and turns the spool with a foot pedal while he winds several loops around the container. When he has wound his "web", he cuts the string with the curved blade of a "spider ring" on his finger. (Some manufacturers don't use string to reinforce a break

The Spiking Room: winding string around a newly star-filled break.

container; they use a heavier container and claim it eliminates the hazard of falling bits of burning string and paper.)

Now the container of stars will have a time fuse inserted into it and be rolled in brown paper, the first of as many as three wrappings before the shell is completed.

Although the Zambellis buy most of their cardboard containers ready-made, Joseph Zambelli can make large cases by rolling heavy kraft paper, as thick as the cardboard in a shoe box, around a form that looks like a big rolling pin. Once he made surprise shells for an Atlanta Braves baseball game fireworks show. Each shell was packed with a Braves T-shirt! Fans were delighted to see the shirts pop out of the shells and float down, unharmed.

Flash Powder Shed: Test Blasts

Putting the "thunder" into fireworks can be even more dangerous than creating the visual razzle dazzle. The big booms you hear are produced by highly explosive flash powder. Mixing flash powder for shells is a job that demands special care, at the Zambelli plant or any other fireworks plant.

On a typical day, one man weighs and mixes small batches of chemicals by hand until he has a total of 1,000 pounds of flash powder. The air becomes filled with gray aluminum dust and the man doing the mixing must wear a face mask and goggles. The metallic dust settles on his hair, clothing and boots until he looks like "The Tin Man" by day's end.

Several times a day, "The Tin Man" tests the power of the chemicals in the flash powder. He takes a one-ounce sample to the Spiking Room where it is poured into a "salute container". The salute container has a much thicker and stronger wall than a star container. More gas will build up before the container bursts, resulting in a more violent, louder explosion.

Co-workers hang the test salute from strings tied between two steel fence posts. (Once a test salute was tied directly to the post and destroyed three inches of steel.) When the fuse

Carefully mixing flash powder, which provides the exciting, thunderous noise of fireworks.

is lit, the man who mixes the flash powder stands far away—with good reason. He's coated with explosive dust and must be on guard to avoid sparks or flames.

The test shell delivers a mighty blast and the flash powder is proclaimed good. Now it can be packed in small salute containers and fitted with a stubby custom-made fuse called a "spaghet". The salute will become the final thunder in the two-break shells under construction in the Spiking Room.

Pasting Room: Rolling in Glue

In the Pasting Room, two women roll star-filled breaks in damp brown paper that has been soaked with a special paste. The paste-soaked paper becomes hard when it dries and gives the break container extra resistance to the explosion. The women tear and mold the paper to fit neatly around the ends of the container and form a tight seal.

For about two days the wrapped containers sit outside drying. If the weather is rainy, they dry inside where heaters provide warm air.

Finishing Room: Final Fuses

As a final step, Louis Zambelli, vice president, and another technician put together all the parts of the shells. They try to complete 500 average-size shells every day.

In Mr. Zambelli's steady hands, separate breaks of stars, gunpowder and different kinds of fuses will be packaged into a multi-break shell whose parts will explode one after the other, exactly as he wants. He and his brother Joseph control the timing of the explosions by the type of fuses they use, their length, how much powder they contain and a built-in "ignition system" that helps light the fuses.

Mr. Zambelli uses his teeth to pull string tight as he ties the fuses in place and seals the outer wrapping. The long starting fuse is folded and tied for easier handling. The last step is labeling the shell so the shooters will know its size and how it will perform in the sky.

Besides the shops where aerial shells are created, the Zambelli manufacturing plant also has rooms for several other activities, including paper cutting, carpentry and making lances for set pieces.

Working with humble bits of string, paper and dusty powders seems worlds away from the glamorous spectacle of a nighttime show. But the quiet, detailed and patient work of craftsmen and women is what makes all that magic in the sky possible.

Newly wrapped breaks from the Pasting Room, drying outside.

Joseph Zambelli tying up a jumbo shell.

	EXOTIC FAN	DOUBLE ROW WHEEL	CRAZY WHEEL	DEVIL WHEEL	LUCE WHEEL
DATE: LOCATION:	FIVE WHEEL	SAXON SQUARE	MAGNET (SLOW)	MAGNET (STEEL)	MAGNET (BRIGHT)
	AIRPLANE	SHIP	CANNON CANDLES	TANK	STICK FINALE
	LIBERTY BELL	CAKE & DATES	SIGN WELCOME	SIGN GOOD NIGHT	SIGN DRIVE SAFELY
COLOR WHEEL	GREEN ELECTRIC	YELLOW ELECTRIC	WHISTLE WHEEL	NIAGRA FALLS WHEEL	PLAY WHEEL
MERRY WHEELS (SILVER)	MERRY WHEELS (GOLD)	SIX POINT STAR	NIAGRA FALLS	GIRANDOLA & STICK	RED WHITE AND BLUE WHEEL
OLD GLORY	INDIAN	COVERED WAGON	DONKEY & RATS	RATS & WIRE	SEAL & BALL

A chart showing some of the many set piece designs a fireworks manufacturer can offer its clients.

Getting Ready

Long before the night of a fireworks show, a fireworks display company begins its preparations. Planning starts with an order from a customer. It could be any kind of company or organization, from a pro baseball team to a county fair committee. Fireworks firms have even been asked to produce shows for private birthday parties and weddings. What type of show will be produced depends on the location, any special theme or purpose the show has, and the amount of money the sponsor can spend. Shows usually cost from $1,000 to $25,000—and a few cost over $200,000. A complex, specially-designed show with original fireworks could take a year's planning.

For most customers, a fireworks expert will recommend one of several standard programs from his catalog. Each would include opening salutes (the big "booms") and a variety of medium-sized aerial display shells. These range from the quiet, spreading, blossom-shaped shells called chrysanthemums, to novelty shells that whistle, hum or chatter like machine guns. A grand finale—a group of shells all fired at once—ends the show dramatically.

For an extra fee, the fireworks specialist has many different effects to offer. He can outline the sponsor's name or symbol in blazing fireworks on a wooden set piece. Or he might highlight a large show with jumbo shells 12 to 16 inches in diameter, whose glare can be seen 25 miles away! He can even time the firing of the shells to coordinate with live or recorded music.

Firing Methods: Flame or Spark?

Depending on the size of the show, the fireworks company will decide whether to light the fuse of each shell by hand with a long flare, or electronically. (Some companies only shoot shows by hand.) Smaller and medium-sized shows can easily be fired by hand. Larger, more expensive shows, or shows that are synchronized to a musical program, are usually fired by attaching wires to each shell's fuse. The paper tubing covering the raw fuse is slit open and a fine wire inserted. Each mortar, or launching tube, is numbered; the numbers are listed on a central switchboard to which all the wires are connected. To fire a shell, the operator simply flips the right numbered switch, sending an electric spark to the fuse. Only one shell per mortar can be shot this way, however, and hundreds of mortars are needed.

For a big show, there can be many, many switches to flip. When Zambelli Internationale presented the annual WKGB Radio Sky Show at a San Diego stadium, it took four control panels with 300 switches each to fire all the shells and set pieces in the program—that's 1,200 switches!

Weeks before any large show, the head of a fireworks display company travels to the show site to meet the sponsors and decide what type of fireworks will be most effective.

Is the show by a lake or river in which fireworks can be reflected? Will the firing site be a stadium or the top of a

 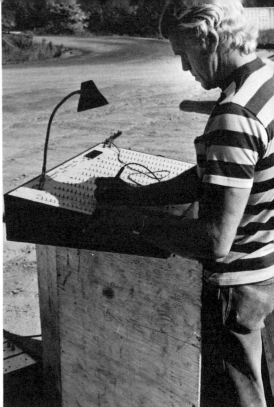

Left: *A fireworks technician attaches electrical wires to shell fuses.* Right: *The other end of each wire is attached to its own switch at this central switchboard.*

building? On grass or a cement-paved area? Each setting requires different firing techniques. The fireworks specialist may also check with local police and fire department officials about safety precautions. Some communities require that a fire truck be standing by during a show.

Designing the Show: First Steps

The fireworks specialist designing a big show combines dynamic elements—brilliant color, sound, patterns, space and time—in almost limitless ways. He can change the color of the sky from all red to blue to white with massive flights of shells. He can pepper a huge, spreading floral pattern with little bursts of single-color star shells. He can treat the sky

For this riverside show, mortars are lined up along the water's edge. When fireworks are launched the reflections on the water nearly double the display's effectiveness.

and the ground as a stage with many levels and shift the action smoothly from ground to sky and back again.

On the ground, he may position set pieces—designs or rocket-spun pinwheels, or fountains or Roman candles. He alternates the ground fireworks with aerial displays. As shells in the sky fade away, a set piece is lighted and blazes for one minute. When there is nothing left but smoke, the audience's eyes are attracted skyward again as a new round of shells reaches its peak and fills the blackness with glitter. The designer also plans combinations of shells, which are fired together, but explode at different heights in the sky. Achieving these effects sometimes takes careful planning at the drawing board.

To the Drawing Board

Well in advance of the show, custom-made set pieces are designed and partially built. The sponsor sends the fireworks company a symbol, name or drawing to be copied. Companies have done fireworks portraits of everybody, from famous football coaches to Mickey Mouse. An artist sketches the de-

An example of what a fireworks show designer can do: Two ordinary comets shot off simultaneously and placed so they criss-cross in the air, create a dramatic, unusual effect. (Antiphonal Comets—Tri-State Fireworks Co.)

sign on graph paper, sectioning the design into squares that equal sections of the completed set piece frame. The drawing is a guide for the carpenter who builds the set piece in small sections that can easily be transported to the show site and nailed together.

Onto each section, the carpenter attaches thin wooden slats in an outline of the design. Then, little colored flares—lances—are stuck on the slats. Just before the show, workers staple a length of fuse to the lances. Fire races along the fuse, lighting hundreds of lances in a few seconds.

The fireworks designer also will draw a blueprint showing the fireworks team where to place the set pieces and the mortars that will launch the shells into the sky. If a background of aerial bursts is supposed to appear behind a glowing set

Left: *Attaching colored lances to a set piece frame.* Right: *Stapling quick-lighting fuses to the lances.*

piece, for instance, the right size mortars must be placed behind the frame. Huge mortars, with mouths 8, 12, or 16 inches wide, are separated by rows of smaller mortars. This gives the big shells plenty of room to spread out in the sky.

Mortars not only have different widths; they come in different lengths. A longer mortar sends a shell higher, just as a rifle shoots a bullet farther than a pistol. Often, the fireworks designer likes to make several larger and smaller shells burst open simultaneously. The larger shells, in longer mortars, are lighted first, because they take longer to climb higher. The smaller, lower-flying shells are launched next. The result is a "stepped" display at various levels.

The display looks richer, too, when shells with differently shaped patterns are shot at the same time from strategically placed mortars. The contrasting shells cluster in the sky like a bouquet of mixed flowers.

"Pyro-Musical" Event

Planning fireworks that will be electronically launched to music makes the job even more challenging. First the show's sponsor sends the fireworks arranger a tape of the music to be played the night of the show.

As the arranger listens, he visualizes what fireworks effects will match the musical effects. If the sound is lush and romantic, he thinks of "soft" fireworks, like the nearly silent Oriental flower shells. If the music is heroic and military, with lots of drums and brass instruments, he pairs it with shells that have plenty of sound *and* color. If he hears the theme from the movie "Star Wars," he may picture rockets which streak through the sky leaving their long tails hovering behind them, and novelty shells that whistle piercingly or that sound like fighter planes battling in the clouds.

The arranger then makes a chart, called a timed count track, listing every second of every minute of the music, and notes exactly when each shell or set piece should be fired. He is guided by the catalog telling him the firing time of each shell. (The average shell takes 15 seconds to climb to its peak height, burst open, burn its "cargo" and fade away.) At four minutes and 50 seconds into the music, for example, he may order a palm tree shell with its long streaming "trunk" and "leafy" branches; at five minutes and 10 seconds, a flight of green star shells, and so on. When the band strikes up "The Star Spangled Banner" the arranger may call for a waving American flag set piece to be lighted, while a flight of many red, white and blue multi-break shells bursts open overhead.

The timed count track enables the operator firing the shells on the night of the show to pull the right switch at just the right time. Usually the show works with amazing precision, but for those rare occasions when an electric igniter fails, fireworks technicians stand by ready to light the problem shells by hand, even though they have to make their way through a field of firing mortars to do so.

Music is only one of the special effects at a fireworks designer's masterful fingertips. Like the ringmaster of a three-ring circus, the fireworks expert may add other spectacles to fireworks because he knows that the more you have to watch, the more dramatic the show will be. He might decorate a hot air balloon with strings of electric bulbs and send it aloft at the climax of the show. Or he could call on the talents of a new partner in nighttime magic: the laser light show artist.

Laser Show on a Smoke Screen

The laser artist, or laserist, creates patterns with laser light beams and projects them up into the sky. The gigantic patterns can be coordinated with the fireworks during a fireworks show and synchronized to the same music.

First developed in 1960 and adapted for artistic light shows in the early 1970's, lasers are machines that create and build up tremendous light energy. The light is allowed to "leak" out of the machine in an extremely brilliant, sometimes powerful, narrow beam that travels long distances. Beams from fairly low-powered lasers have been bounced off the moon and stayed bright enough to be visible back on earth.

For most laser light shows a krypton gas laser is used. This produces a rainbow of colors—red, blue, green and yellow. The colors are separated by a prism and the beams are bounced off two small mirrors. The mirrors can be vibrated to make the light beam "dance" to any music, from a romantic waltz to a pounding rock rhythm. Also, special lenses can be placed in the path of the beam to create webs and whirling loops that are like the patterns you create with a Spirograph® drawing set. The laserist can mix colors, move the light patterns and change the brightness of the beam.

At the closing of the 1982 World's Fair, the Zambelli Internationale show included a laser which beamed the symbol of the fair into the sky, along with other patterns. The laserist had to follow the fireworks activity closely to aim the beam at the same altitude as the exploding shells. He also adjusted the

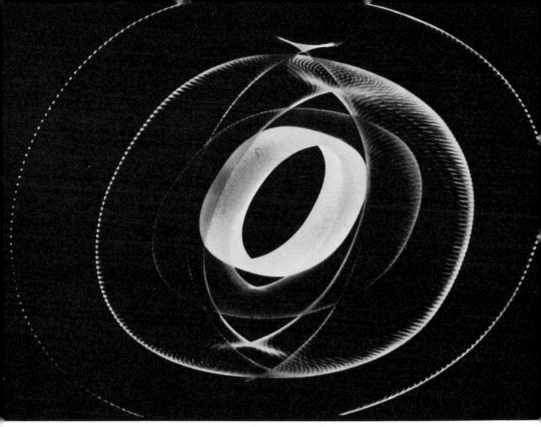

A laser light pattern, courtesy of Zambelli Internationale.

colors and brightness of the beam to contrast with the ever-changing fireworks.

The smoke in the air from a fireworks show creates a good outdoor screen for a laser light show. The smoke scatters the laser light beam, making its pattern clearly visible in the sky. Laser light seems to sparkle as it pierces the smoke.

Laser light shows and fireworks may be a natural combination for nighttime entertainment, but good weather for fireworks is not good weather for laser beams. "You need a good clear night for fireworks," says George Zambelli, "but you can see laser better on a hazy, overcast night when the smoke hangs in the air." If it is too windy, the smoke vanishes too quickly.

Using lasers in a fireworks show can add several thousand dollars to the bill. Not only are special equipment and technicians needed, but extra fireworks material must be fired to maintain a heavy smoke screen. When a show marks a special event, though, lasers are an exciting extra dimension.

Packing Up and Pulling Out

Packing is a big job. A spectacular, half-hour show can take as much as ten tons of fireworks and equipment, and in the hectic Fourth of July season, dozens of truckloads must be sent to hundreds of shows. Loading often goes on late into the night.

Each shell for an electrically fired show will be marked with its mortar number. Shells for hand-fired shows will be packed in special wooden boxes—or ammunition "depots"—in the order in which they will be used. Although the shells are labeled, the fireworks team will be working in near darkness. They depend on a well-organized packing job to help them locate the shell they want.

The Show Team

Next, the fireworks company selects the crew that will actually put on the show. The basic fireworks display team includes a *loader*, who loads shells into their mortars, and a *shooter*, who ignites the shells' fuses by hand with a burning flare. In a large electronically fired show, the whole team acts as loaders before showtime. Only one shell will be shot out of each mortar, so hundreds of mortars must be pre-loaded with shells and the shells wired to the electrical system.

Loaders and shooters are trained on the job or at fireworks "schools" offered by some fireworks companies. In a fireworks family, the older generation teaches its sons and daughters and nephews and nieces the art. But with hundreds of shows to shoot every July Fourth, a fireworks company needs more help than family or full-time workers can provide. Each year it must call in an army of trained "moonlighters," or part-timers. At Zambelli Internationale, for instance, the regular staff of about 50 people swells to 1,700 people to meet the Fourth of July rush.

These moonlighters are not amateurs. Many are licensed fireworks technicians who can run a show on their own. In states with strict licensing regulations, such as South Carolina and California, a person can't qualify for a license as a fireworks operator or assist in a fireworks show until the age of 21. In South Carolina, a worker must have shot fireworks in at least six shows, have a letter of recommendation from a display company and pass a written test before he or she can be certified.

Each large fireworks company keeps a list of trusted, experienced technicians who live around the United States. When a company in the northeast gets a job in Atlanta but doesn't have enough full-time workers to send there, it can hire fireworks experts who live in Atlanta. Or it might hire local part-timers and fly them to Atlanta.

Fireworks moonlighters are well paid for their skill, although many shoot fireworks for the fun of it, too. They may be college students, factory workers, fire fighters, businessmen, hotel or amusement park managers, doctors or nurses. One summertime shooter for Southern International Fireworks in Rock Hill, South Carolina was a sixth grade teacher. He carried his camera with him constantly and shot many rolls of film during the months he was shooting fireworks. When school opened, the teacher presented a unique slide show to his sixth grade class on "What I did this summer". He says his class was never so attentive and quiet.

Fireworks Tonight!

It is the morning of a fireworks show. Gary Partlow, president of Southern International, is checking the equipment he will bring to the show site thirty miles away. At the same time he is preparing himself mentally for the night's exciting—but hazardous—work. Although Gary teaches fireworks shooting around the country and has carefully drilled his own crew in safety procedures, he is always nervous before a show. "If you're not," he says, "you shouldn't be in the business. Shooting fireworks *is* dangerous. It's never out of your mind."

Before taking off for the show site, Gary and his team take many precautions, packing fire extinguishers, and a first aid kit. An extra crew member will be recruited as a "spare" in case one of the team gets hurt. Safety isn't the only thing Gary is concerned about. Equipment must be checked and double-checked to make sure the team will have all the shells, mortars and tools they'll need. Gary scans the sky for rain clouds and listens to the latest weather forecast. To help his team put on a beautiful, smooth-working show, he goes over the night's plans several times with everyone. He wants their timing and teamwork to be perfect.

This July Fourth weekend show must run exactly 17 minutes, the same length as a musical tape that will be played at the show site.

Gary's wife Kathy will be the leader for tonight. She will time the show with a stopwatch, and tell the crew when to fire the most spectacular shells like the eight-inch butterfly shell from China that spreads sparkling wings. She also helps pace the firing of the smaller shells.

Jesse "Marty" Matthews, a fireworks technician whom Gary has worked with for seven years, will act as a shooter tonight. At this show, which will be hand-fired, Marty will be the leader of one firing team. He helps decide where mortars are to be placed on the site. And, once the show begins, he calls out to the loader what type of shell to load, and lights the fuses that send each shell skyward.

In terms of safety, the critical job is the loader's. Barbara, Marty's wife, is a loader for this show. She will be responsible for placing shells of varying diameters into the right size mortars. If she accidentally puts a four-inch shell into a five-inch wide mortar, gas from the lifting charge at the base of the shell escapes and won't exert enough pressure to propel the shell upward to a safe height. A powerful shell that explodes low to the ground can seriously injure a member of the fireworks team. Gary has a partial hearing loss from a shell that exploded just over his head. He doesn't know whether the shell misfired or the apprentice loader he was training put the shell in the wrong size mortar. In a fireworks show in which hundreds of shells are fired by hand, Barbara must work quickly and accurately. Everyone's safety depends on the loader.

Setting Up

Gary and his crew arrive on the site at 6:00 p.m. The show will start at 9:30. Their first job is setting up the mortars. Gary

A crew member covers the loaded mortars with aluminum foil, to be sure the shells stay dry.

has to be sure the mortars are aimed at an angle that will prevent burning shell casings from falling down on the spectators, their cars, and nearby buildings. He fires a test shot to check the direction of flight and the effect of the wind.

The team starts digging holes in the ground for the biggest of the mortars. When they are done they will lower these mortars into the holes so that only a foot of each launching tube is above ground level. Burying the mortars this way gives the crew extra protection in case a shell explodes before leaving its mortar. When holes can't be dug, large mortars are set up in sand-filled wooden frames or 50-gallon steel drums filled with sand and surrounded by sand bags. Smaller mortars are set up in wooden racks.

To prevent mistakes and help organize the firing, the crew

carefully groups mortars according to the size shell that will be fired from each. Matching shells are located directly behind them in a "depot" box. A team of shooter and loader will be assigned to each size mortar group.

In a hand-fired show like this one, Barbara loads each mortar again and again: as many as 15 three-inch shells will be fired from a single mortar. Only iron and steel mortars can stand the heat the repeated firings produce. For an electronically fired show, heavy cardboard mortars are used because each mortar is used to fire only a single shell.

While the mortars are being readied, other members of the firing crew are erecting set pieces. They pound wooden posts into the ground and mount stars and pinwheels on them. Larger set pieces are put together and fused on the ground before being hoisted into place.

It is a good, clear night. If the weather were threatening rain, the crew would cover the mortars with aluminum foil; the shells shoot right through the foil. They would also probably drape the set pieces with heavy plastic until showtime,

A lit pin-wheel set piece, reflected in the water.

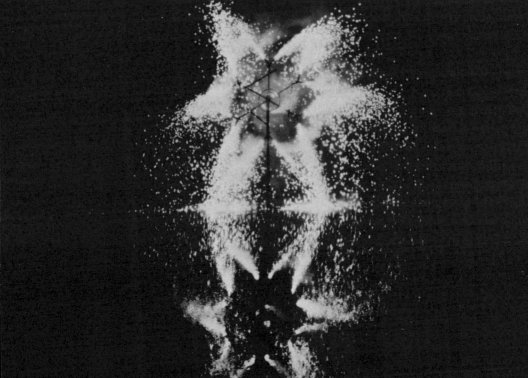

The same set piece, before show time.

to make sure all the fuses stay dry. A light rain won't cancel the show. In fact, sometimes a drizzle gives the fireworks a beautifully hazy look.

It can take up to three days for a large electronically controlled show to be set up. Even a small hand-fired show that will last just 15 minutes requires a few hours' preparation. By showtime, Gary and his crew are as eager as the audience.

Showtime!

Around nine o'clock there is a last minute flurry of activity. Boxes of fireworks shells are quickly unloaded from trucks where they have been safe from curious bystanders and sudden showers. Mortars are loaded with the first round of shells. One hundred and fifty shells are placed in the zig-zagging finale racks, fused together and covered with aluminum foil, in case the weather changes. As the area clears of vans and cars, and friends and families of the crew leave for a safe watching spot, Gary gathers his people together for a final briefing.

Just a little earlier some crew members were laughing and joking about "shaky hands". A couple even showed off funny firing techniques, twirling and leaping away from imaginary fuses. As Gary talks, however, everyone grows serious. Faces look tense. A dangerous and difficult job is ahead.

The crew listens intently as Gary assigns shooters and loaders to different size mortars. Marty and Barbara will take the biggest mortars. Flares are handed out to light the shells' fuses. Electric lanterns are passed around to shed a little light in the firing area. Again, Gary runs through the show's schedule. He issues a few last instructions and warnings. "Don't shoot when the shooter beside you shoots. Alternate. When Kathy calls out that we've reached the 15 minute point in the show, I want you to get all the shells out of the boxes—fast. Get all the mortars loaded. Don't watch the sky. Keep your eyes on your job down here." The crew members disperse to take their positions, ready to start.

As the sky grows completely dark, the crowd's attention is drawn by the firing of a few salutes. These are sharp cracks of sound and little more than smoke. But soon the sky is the backdrop for a magnificent variety of shimmering colors and patterns. Gary and his team keep the sky ablaze nearly the whole time.

Down below, the firing site is like an artillery range on a battlefield. Clouds of smoke rise above the mouths of the mortars. The smell of gunpowder and other pungent chemicals fills the air. After each explosion, burning fragments of paper casings and string drift down. The more elaborate the shell—the more breaks—the more debris descends.

If the air is still or the wind is blowing against the shooters and loaders, the burning bits can land on them. Their shirts frequently get "air conditioned" with burn holes and their skin gets the same kind of minor burns that welders get from sparks. During a show, just as at the factory, fireworkers

The firing site: noise, smoke, the shooters' bright flares, watchful team members standing by.

wear cotton shirts and pants, never synthetic fabrics that are more flammable. On a hot summer's night, like tonight, the workers' shirts are soaked with sweat from hard work and the intense heat from the firings.

Midway through the show the wind picks up and shifts slightly. Debris in the air from the large mortars is blown onto the men firing the three-inch shells. There is nothing they can do to avoid the heavy rain of burning bits. Like soldiers, the two loaders and shooter stick to their work—except for one scarey second when, suddenly, a large, flaming paper casing sails down toward them. Someone cries a warning. The men instantly jump aside. The casing hits the ground and one man stamps out the flames. Watching the action, Gary wishes he had placed the three-inch mortars ahead of the rest of the firing line.

Through all the noise and unpredictable falling of burning paper and string, Gary and his team are watching out for each other's safety. Every time Barbara takes a shell from the "depot", she must be sure to cover it again with its wooden or canvas cover. If a spark or bit of burning paper landed on the shells, it might set off a tragic explosion. Another major hazard, particularly in a hand-fired show, is the shell that shoots out of its mortar and lands in the midst of the fireworks crew. Fast work with the ever-handy fire extinguisher may prevent injury or, if there are no seconds to spare, the crew may have to make a run for safety.

One shell is lighted but nothing happens. Barbara will remember where this "dud" is sitting and not re-load its mortar. If a shell is smoldering, she will grab the fire extinguisher and spray it. Both Barbara and Marty pat out sparks that land on each other. They also call out the location of any shells that

Two team members, a shooter and a loader, work to keep the sky ablaze.

do not explode properly in the sky. These they will take care of later. Otherwise, the fragments are sitting time bombs that could seriously injure an unsuspecting souvenir hunter.

Firing a show requires other precautions, too. Prolonged exposure to high noise levels could damage a crew member's hearing. Marty, however, like many fireworks technicians, doesn't like to wear ear plugs. He does put them on, reluctantly, during the busy Fourth of July season, or when he is shooting eight-inch or larger shells. James Sorgi of American Fireworks Company has his own method: he turns his back after lighting shells and folds his ears forward "so the sound hits the back of my ears". After a lifetime in the fireworks business, he says his hearing is fine.

Throughout the show, Kathy Partlow, stop watch in hand, quickly paces among the scurrying crew members. Every few minutes, between the blasts and booms, she yells out how many of the show's 17 minutes have passed.

At five minutes into the show Kathy cues Marty to send up the big butterfly shell. At eight minutes she calls, "Palm trees. Everybody hold your fire." Three palm tree shells blast out of their mortars and crisscross in the sky. Soon Kathy is shouting, "Fifteen minutes. Fifteen minutes. Get everything out."

As the stirring music of the "1812 Overture" rings out, the sky fills with a mixture of special effect shells and multibreaks, rockets, comets and giant flowers. Then the finale racks send up a flood of shells in an ear-pounding, dazzling crescendo of sound and color that seems to double, triple and quadruple before it ends.

The audience cheers, applauds and honks its car horns. It's time to press homeward, through human traffic jams and clogged streets, since the show was a popular one. The action isn't over, however, for Gary and the crew.

There is clean up work to be done. Hot metal mortars must be carefully pulled from the ground or their racks and rolled toward the trucks. Paper casings are collected and thrown away. Those suspicious "duds" that didn't explode properly have to be found before they cause an accident. Set pieces must be taken down and dismantled. The crew wants to pack all their gear in the trucks and take it directly to the site of the next night's July Fourth show—the top of a nearby office tower. After that, the tired crew will head out for a late dinner to celebrate a job well done.

For Gary's crew, or any fireworks team, the show is far more than a job. Like dedicated performers, the technicians are proud when a presentation comes off perfectly—and thrilled when the audience responds warmly.

A finale.

An 1867 drawing of a child playing recklessly with firecrackers.

Fascinating Fireworks Facts

Worst Fireworks Disaster

Sometimes the huge crowds attracted to fireworks shows cause more safety problems than the fireworks themselves. On May 16, 1770, the citizens of Paris packed into a large public square to watch a grand fireworks display in honor of the marriage of their future king, Louis XVI, and Marie Antoinette. When the show was over, however, the crowd could hardly move. Panic-stricken, people began to push and shove. Many were knocked down, crushed and trampled as the crowd surged out of control. More than 800 died and many were injured. In sympathy, the royal couple gave money to the victims but the public was not pleased.

Living Fireworks

Imagine these scenes acted by performers "wearing" fireworks: Jack climbing the beanstalk; firemen unrolling a hose and spraying a burning house; a village blacksmith hammering sparks from hot metal; a boxing kangaroo; whale hunters harpooning whales; and fighting bears. These were typical Living Fireworks Dramas presented by C. T. Brock at London's Crystal Palace in the 1880's and 1890's.

How did the performers work without being burned? Mr. Brock invented a way for the performers, dressed in fireproof asbestos cloth overalls, to wear a light wooden frame on the side of their bodies closest to the audience. The frame was covered with colored lances outlining the shape of the character or animal portrayed.

The living figures acted before elaborate set pieces that served as stage scenery. For the firemen scene, a burning house would be the backdrop. For "Life in the Arctic Regions," the set piece was aglow with spouting whales in a sea filled with icebergs.

Sometimes the Crystal Palace shows featured a blazing tight rope walker who made a daring slide to the ground.

"Atomic" Fireworks

When the U.S. government wanted a simulated (fake) atomic bomb for training soldiers in the 1950's, Felix Grucci, Sr. pictured in his mind exactly what was needed. He mixed some chemicals that he thought would create a heavy white smoke. When he tried his "bomb," the results were better than he expected. His A-bomb had a towering stem and a huge mushroom cloud of smoke which hung in the air for five minutes. Deadly looking, but safe!

America's Last Gunpowder Mill

Old fashioned blackpowder or gunpowder is a key ingredient in fireworks but there is only one plant in America that makes it. Goex, Inc. (formerly an E. I. Du Pont de Nemours & Co. plant) in Moosic, Pennsylvania mixes and grinds a variety of fine and coarse powders for the fireworks industry. Once there were many blackpowder mills to keep our guns and cannons loaded. Today, blackpowder has largely been replaced by smokeless powder, a different chemical compound.

Fireworks at the Movies

"The Wizard of Oz" (1939) was one of the first movie musicals in color and the first to use colored fireworks for special effects. Billowing red "smoke" in the Wizard's throne room created an eerie atmosphere. Fireworks have been used in many movies since. In the 1981 British fantasy film "Time Bandits," a pyrotechnist is listed in the film's credits. Many simulated bombs were exploded in the historical war scenes. "Star Trek II: The Wrath of Khan" also used the talents of a fireworks specialist. The Gruccis provided fireworks for scenes in "Godfather II," "Blow Out" and "Manhattan."

Fireworks on the Moon

In the space age, pyrotechnics mean more than traditional fireworks. Pyrotechnics are also a group of devices using explosive, propellant (moving another object) and pyrotechnic compositions to do special jobs, even on the moon. Pyrotechnic cartridges on an airplane might be used in an emergency to blow off a hatch or door so passengers and crew can escape. Pyrotechnic devices, with their self-contained energy, can back up another energy system—like an electrical system—if it fails.

Each Apollo mission to the moon had at least 310 pyrotechnic devices on board to perform controlled tasks, some vital to the safety of the crew. For instance, a pyrotechnic thumping machine was placed on the moon by astronauts to create mini-moonquakes. Scientists could measure the vibrations to learn more about the moon.

Buried with a Bang

Antonio Nerti, a West Pittsburgh steelworker, didn't want the usual funeral. Instead, in his will he asked for a marching band and a five-minute Zambelli fireworks display and left $750 for his final tribute. The funeral in 1977 would have made him happy. As the grand finale, 300 rocket-launched flags fluttered to earth on little parachutes.

World's Biggest Wheel of Fireworks

The 1981 wedding of Prince Charles of England to Lady Diana was celebrated in traditional style—all the way from fireworks to the coach ride to the church. On the night before the wedding, a London park was the scene of one of the largest fireworks displays of its kind staged in England in over 200 years. A gigantic fireworks wheel, 40 feet across, shot fire over a 100 foot area as its rockets spun it around. The Household Musicians, in red coats and beaver hats, and two choirs accompanied the lighting of a palace set piece decorated with the Prince's insignia, the three ostrich plumes of the Prince of Wales.

World's Biggest Dud

Before there was record-breaking Fatman II, the world's largest firework, there was Fatman I, the world's largest dud. Fatman I was a test shell built in 1976 by the Grucci family for $1,500 for the writer George Plimpton, New York Fireworks Commissioner and a serious fireworks fan. If it worked, a similar shell would be fired at Harvard University.

Alas, the 720-pound Fatman I never left its 40-inch wide mortar buried in the sand of Westhampton Beach, Long Island. The whole aerial shell, including hundreds of white stars, burst underground. The lifting charge had failed to ignite properly. When the smoke cleared, there was a crater big enough for a house foundation.

A Million Dollar Fireworks Show

Each year, a Houston, Texas department store, Sakowitz,

advertises fabulous "Ultimate Gifts" in its Christmas catalog. In 1981, one of the gifts was a one million dollar fireworks show custom-designed by Fireworks by Grucci. The 45- to 60-minute performance included 25,000 aerial shells, high-intensity spotlights, lasers and fireworks amid "dancing waters" fountains. Everything would be electronically synchronized to original music specially composed for the occasion. George Plimpton, the writer, would narrate the show in person. It would take the Gruccis six months to create the spectacle. No one ordered the show but it attracted lots of publicity for the store and the Gruccis—just as they hoped.

Fireworks of a Different Color . . .

Throughout the centuries, fireworks have been used in some surprising ways, from keeping hungry birds off farmers' crops, to helping stranded motorists attract attention. Indeed, fireworks are such good attention-getters that policemen often use hand-held flares to direct traffic at accident scenes. Fireworks can also be used to help mark a sinking ship's position. Ships' captains are required to carry emergency kits which usually include an assortment of colored flares and a rocket launcher.

HOW TO WATCH A FIREWORKS SHOW SAFELY AND HOW TO FIRE HOME FIREWORKS SAFELY

1. Stay in the designated watching area far away from the front and rear of the fireworks display area. Remember that burning shell fragments, "duds," and shells that explode too close to the ground make any place near the firing area a dangerous place to be. You will be able to observe the full effect of the aerial shells best at 500 feet or more from the display area.

2. Never pick up or take home a "souvenir" shell you find after the show.

Even if it is only a fragment of a shell, it may contain a large amount of explosives. The slightest spark or shock (such as dropping it) could set it off. Call a police officer or firefighter and tell them where you found the shell.

3. Never enter the display area before or after the show looking for shells. Following the show, there may be unexploded shells in some mortars. Fireworks technicians know how to dispose of them safely, but you don't. Shells this large could kill or badly injure people if they explode at ground level.

4. If a laser light show is part of a fireworks show, the Federal Food and Drug Administration suggests keeping these safety rules in mind. Never look directly at the source of a

laser beam. Never look at a shiny surface that is reflecting laser light. Never look at a laser source through a camera viewfinder or binoculars. Look only at the reflected light pattern in the sky. Laser light can seriously damage your eyes, but you won't know it until you notice a loss of vision. The laser can painlessly burn your eye's retina.

If you live in a state where home fireworks are permitted and you decide to set some off, the Consumer Product Safety Commission recommends following these safety guidelines:

ALWAYS read directions. If you don't understand the instructions, don't take a chance on lighting the firework.

HAVE an adult present. Proper supervision can prevent dangerous mishandling.

NEVER experiment. Don't take fireworks apart or mix anything with their contents. Never try to make fireworks yourself.

IGNITE outdoors. Light fireworks in a clear area, away from houses and away from all flammable materials (dry bushes, gasoline cans, etc.)

LIGHT only one at a time. And move back quickly once the fuse catches fire.

HAVE water handy. Keep a bucket of water nearby for emergencies and for dousing misfired fireworks.

KEEP a safe distance. Be sure others are out of range before lighting any fireworks.

DISPOSE of properly. Do not try to relight or handle malfunctioning fireworks. Soak them with water and throw them away.

NEVER give fireworks to small children. They are too young to understand the dangers involved.

STORE in a dry, cool place if you won't be using them immediately. Check instructions for storage of special types. Avoid rough handling that might damage fuses or handles.

ALLOW enough room for proper function. *Never* ignite fireworks in metal or glass containers.

POPULAR HOME FIREWORKS

Each state has its own laws about the kind of Class C home fireworks you can buy. Which of these fireworks, if any, are sold in your state?

NON-EXPLOSIVE

Smoke Balls: Small balls of different chemicals that, when lit, give off smoke of a single color—often blue, green or orange.

Snake: A small pellet of chemicals wrapped in cardboard or foil. It burns slowly and produces a long trail of ash, over a foot long. The ash "crawls" out of the container like a snake.

Sparkler: A steel wire coated at the top with spark-producing chemicals. When lighted, the chemicals burn at a high temperature and throw out sparks as far as six inches. Sparks may be red, white, green or gold.

Aerial Shells

EXPLOSIVE

Some of these fireworks are smaller, simpler versions of larger fireworks fired by professionals.

Aerial Shell: A wrapped container of explosive powder and color or spark-producing materials that is shot out of a cardboard mortar. Class C shells are the "little brothers" of the giant shells fired by professionals. They contain only 130 milligrams of powder each. They may contain multiple "breaks", producing changes in color or a final loud report.

Fountains

Firecracker: A slender paper tube (about 1½ inches long) containing no more than 50 milligrams of explosive powder. When the powder burns, gas is formed that pushes out at the walls of the tube. The tube bursts open with a loud noise. Firecrackers are sometimes strung together so they explode one after the other. Small firecrackers are called ladyfingers.

Fountain: A cone or tube filled with color and spark-producing chemicals. The tube is mounted on a wooden or plastic base so it will not tip over. The cone sits on its widest end. Sparks and burning matter are forced out of a small opening at the top of the tube or cone, somewhat like a volcano erupting.

Fountains

Rocket: A short tube, usually two inches, attached to a long stick. Sometimes the tube is topped with a cone to stabilize its flight. As explosive powder in the tube burns, gases escape from a hole or choke at the bottom of the tube, pushing the rocket up into the air. A second explosion in the air creates a loud report. Rockets also scatter brilliant trails of colored sparks.

Rockets

Roman Candle: A long tube filled with several color stars separated by layers of special explosive powder. The tube is stuck in the ground and lighted. When each layer of powder explodes, it shoots a star 12 to 15 feet into the air. Before each star is launched, a low fountain of sparks flies out of the tube.

Novelty Fireworks: Small ground or aerial fireworks using rockets (without sticks) and color and spark-producing chemicals to create special effects. They include toy tanks, ships, buzz bomber planes, ground "flowers", and spinning tops and wheels.

Roman Candles

ILLEGAL FIREWORKS

These fireworks were banned in 1966, but are still sold. They all contain a dangerously large amount of flash powder. They cause most of today's serious fireworks injuries.

Cherry Bomb: A small red ball about an inch wide.
M-80: Usually a red cylinder, about two inches long, with a fuse sticking out of its side.
Silver Salute: A silver cylinder, about one and a half inches long, with a fuse coming out of its side.

If you buy fireworks only from a fireworks store which sells Class C materials, you will avoid the danger of illegal fireworks. Don't buy fireworks in the street.

* * *

If you don't already know your state law on fireworks use, you can look it up on this list:

These states have outlawed the sale and use of all Class C (home) fireworks:
Arizona, Connecticut, Delaware, Georgia, Massachusetts, Minnesota, New Hampshire, New Jersey, New York, North Carolina, Ohio, Rhode Island, Vermont, West Virginia

These states have outlawed all home fireworks except sparklers:
Colorado, Florida, Illinois, Maine, Maryland, Pennsylvania, Utah

These states have outlawed all home fireworks except sparklers and snakes:
Iowa, Oregon, Wisconsin

These states allow home fireworks, subject to local laws:
California, District of Columbia, Idaho, Indiana, Kansas, Kentucky, Michigan, Montana, Nebraska, New Mexico, North Dakota, Oklahoma, South Carolina, Texas, Virginia, Washington, Wyoming

These states allow all home fireworks:
Alabama, Alaska, Arkansas, Louisiana, Mississippi, Missouri, South Dakota, Tennessee

These states have no home fireworks laws:
Hawaii, Nevada

A SPECTATOR'S GUIDE TO FIREWORKS

At your next fireworks show, see which of these aerial shells you can identify in the sky. Shells with a round pattern spreading out from a center burst are usually Oriental. American or European shells form uneven patterns. Their colors are deeper and last longer. They may be multi-break shells that produce a series of bursts at different points in the sky. Of course there are many shells you barely see, such as the noise-making salutes or the machine gun flashes of "battle in the clouds" shells.

A **Chrysanthemum shell** bursts open to a closely-knit pattern of stars which holds its round shape before fading. It may change color or have a center or outer ring of another color. Other variations are possible.

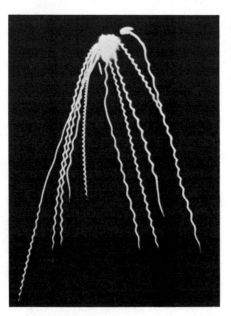

A **Weeping Willow Shell** resembles the drooping branches of a willow tree. The "branches" of the display are streams of color-producing stars which contain charcoal. The charcoal usually produces an orange "fire" and lasts for a very long time—much like the charcoal fire used in an outdoor grill.

A **Peony shell** forms a loose pattern of color stars which breaks up and droops downward. A peony may have an additional center break of streaming "pistils" similar to the pollen-bearing stalks in the middle of a real flower.

A **Palm tree shell** is basically a comet shell with a tail that burns on the way up. Silver or gold "flitter" (metallic powders) give the palm's branches and tail their feathery look.

A **Spider web shell** has a far-spreading web of gold flitter and is American-made.

Rockets have a burning trail of sparks shooting straight down. There is no burst from the top of the rocket; it is propelled upward by gases escaping from its bottom. Sparks fly out with the gases.

Hummer or **Whistle Shells,** above, get their name from the sound produced by gas escaping through a special tunnel-like opening in the shell. The pitch of the whistle, like any wind instrument, can be "tuned" by the fireworks maker.

Red-Tipped Comet, below, left, opens with a ring of brilliant red stars, which create its dazzling red tips. These are followed by streams of flitter. **Finale shells,** below, right, shot by the dozens, provide a dramatic finish to a show.

Selected Bibliography

BOOKS

Brock, Alan St. Hill. *Pyrotechnics*. London, England: Daniel O'Connor, 1922.

Hecht, Jeff and Teresi, Dick. *Laser: Supertool of the 1980's*. New Haven, Conn.: Ticknor & Fields, 1982.

Pain, Henry J. *Pain's Fireworks in Many Lands*. 1921

ARTICLES

Babcock, Martha K. "How do the Zambellis light up the skies so brightly on the Fourth of July? Very Carefully." *People*, July 7, 1980.

Boatner, E. B. "Ancient Art of Chemical Fire." *SciQuest*, July 1979.

Bongartz, Roy Jr. "Bombs Bursting in Air." *Popular Mechanics*. July 1980.

Low, Betty-Bright. "Pyrotechnic Paeans Have Been Flying High for 600 Years." *Smithsonian*, July 1980.

Plimpton, George. "And the Rockets Red Glare." *Sports Illustrated*, June 30, 1975.

———. "First Family of Fireworks." *New York Times Magazine*, June 29, 1980.

Index

A History of Fireworks, Brock, 41
Accidents, 16, 29, 36-38, 43, 61, 65, 101, 104; damage from, 60, *61*, 62. *See also* deaths; injuries; safety
Adams, John, 10
Aerial display shells, 31, 41, 55, 56, *56*, *57*, 58, 60, *112-115*; multi-break shells, 58-59, 66, *75*; manufacturing of, 65-73; strobe shells, 45, 108. *See also* breaks; fireworks; stars
American Bicentennial, 16, 38
American Fireworks Company, 46, 97
American Pyrotechnic Association, 15, 34, 68
Apollo (space mission), 103
Asian style fireworks, 66
Atomic fireworks, 102

Bacon, Roger, 23, *24*
Bate, John, 52, *54*
Blow Out, 103
Boston Pops Orchestra, 44
Bouquet of Chrysanthemums Hanabi, *42*, 43
breaks, 58, 59, 70, 74, *75*
Brock family, 39; Alan St. Hill, 41; Arthur, 41; Charles Thomas, 39, 102; shows, 39
Brooklyn Bridge: use in fireworks displays, 41; centennial celebration, 45
Bull of Fire, 38

bursting charge. *See* charges
Buzz Bombs, 14

Catholic Church, The, 25
charges, 55; bursting, 55, 58-59, 66, 71; lifting, 55; sound, 55, 59, 60. *See also* gunpowder
Charles V, 24
cherry bomb, 16, 110
Chicago Tribune, The, 12
Chinese: contribution to fireworks, 9, 18, 21, 23
chrysanthemums, 77, *112*, *57*. *See also* aerial display shells
Conkling, Dr. John, 68
Consumer Product Safety Commission, 15, 18, 107
Crusades, 23
Crystal Palace (London), 30, 102

deaths, 10, 29, 43, 61, 65, 101; in the United States, 11-12
Death's Busy Day, 10
Democratic National Convention, 48
Democratic Party, 48
Dominion Day (Canada), 46
dragons, *19*, 27, 28
Du Pont de Nemours and Company. *See* Goex Inc.
dynamite, 10

Eiffel Tower, 46
Elizabeth, daughter of King James I, 27
England, 30, 36
Europe. *See* fireworks
European style fireworks, 65

Fatman I, 104
Fatman II, 60, 104
Federal Bureau of Alcohol, Tobacco, and Firearms, 16, 18
Federal Food and Drug Administration, 106
finale shells, *115*
Fire Marshalls Association of North America, 18
fire pictures, 30
firecrackers, 9, 15, 19, 21, 23, 51, 108
fireworks: in America (The United States), 10-12, 31, 32, 65, 66; Asian style, 66; banning of, 14, 16, 18, 110-111; celebrations, 18, *19*, 22-24, *26*, 27, 32, 36, 41, 61, 77, 101, 104 (*see also* Fourth of July); chemistry of, 52, 55-56, 58-59, 60; classification of, 15, 108, 110; coloring agents, 29, 38; competition within industry, 34, 35; cost of, 19, 25, 34, 44, 45, 77; danger of, 12, 16, 18, 23, 26, 29, 62, 72, 89; displays, 23, 25, 26 (*see also* shows); electronically controlled, 30, 50, 78, *79*, 83, 86, 93; in Europe, 23, 25, 26, 30, 35, 36; European style, 65; families, 33, 34 (*see also* specific family names); history of, 10, 21-32; illegal manufacturing and selling of, 16, 18, 61, 110; ingredients, 21, 29, 38, 72 (*see also* gunpowder); largest, 60, 104; manufacturing of, 64, 67-74; in movies, 103; sale of, 14, 16, 18; technicians, 31, 86, 87; testing of, 15, 72, 73; types of, 14, *14, 15,* 16, 25, 27, 31, 43, 51, *53, 76,* 108-110, *112-115*. *See also* accidents; laws; safety; shows
Fireworks by Grucci. *See* New York Pyrotechnic Products
Fireworks Commissioner of New York, George Plimpton, 43, 104
Fireworks Suite (Handel), 36
flash powder, 16, 58, 72, 73
flowers of fire. *See* Hanabi
fountains, 25, 41, 108
Fourth of July, *8,* 10-12, *11,* 32, 39, 44, 46, 86-87 *passim,* 97
France, 25-26, 29, 35-36
friction, 29, 68
fuses, 15; primary, 59; spaghet, 73; starting, 74; time, 24, 59, 72

George II, 36
girandole, 31
Godfather II, 44, 103
Goex Inc., 102
Green Men (Wild Men), 27, *54*
Green Park, London, 36
Grucci family, 44, 45, 60, 103-105; Felix Sr., 44, 101; events supplied fireworks for, 44; Fireworks by Grucci, 104; record holders, 60, 104
gunpowder, 9, 21, 22, 23, 51; burning rate, 55; classification of, 52; ingredients, 21, 52; in weapons, 22-25; mill, 102; use in fireworks, 52, 55, 71

Handel, George Frederic, 36
Hanabi (flowers of fire), 43
Hearst, William Randolph, 61, 62
Holy Roman Empire, 24
hummer shell, *115*

illuminations, 10
illegal fireworks. *See* fireworks
injuries, 10, 12, 16, 19; cause of most, 103; hearing loss, 90, 97. *See also* accidents
Independence Day. *See* Fourth of July
international fireworks competition, Monte Carlo, 44
International Freedom Festival, 46
isinglass, 27
Italy, 25, 48

Jack-in-the-Box, 19, 51
Jefferson, Thomas, 39
"Jimmy the Bomb." *See* Sorgi, Vincenzo
Johnson, Lyndon B., 48

Knoxville, Tennessee. *See* World's Fair
krypton gas laser, 84

Ladies Home Journal, 12
Lake Placid, 45
lances, 31, 39, 81
laser shows, 84-86, *85;* safety rules, 106, 107
laws, 10, 63, 108, 110, 111. *See also* safety regulations
living fireworks, 101
lifting charge. *See* charges
loader, 86, 90, *96*
lockjaw, 11
Louis XIV, 25
Louis XV, 36

M-80, 16, 110
Macy's, 43, 44
Madison Square Park, 61

118

Manhattan, 103
maroons, 55, 56
marrons, 56
Marutamaya Ogatsu Fireworks Company, 43-44
Matthews, Barbara and Marty, 90-97 *passim*
Monte Carlo international fireworks competition, 44
mortars, 58, 59, 78, *80*, 82, 91
Moslem soldiers, 23
movies, *Blow Out*, 103; *Godfather II*, 44, 103; *Manhattan*, 103; *Star Trek II: The Wrath of Khan*, 103; *Time Bandits*, 103; *Wizard of Oz, The*, 103

Napoleon I, 29
Napoleon III, 29
NASA space program, 46
National Boy Scout Jamboree, 50
National Society for the Prevention of Blindness, 12, 18
Nerti, Antonio, 103
New York Mets, 44
New York Pyrotechnic Products Company (Fireworks by Grucci), 44, 60, 105
Niagara Falls, 41
Novelty fireworks, 109

Ogatsu family, 43; accidents, 43; Bouquet of Chrysanthemums Hanabi, *42*, 43; Kyosuke, *42;* Toshio, 43
Olympics, 1980 Winter, 45
Orange Bowl, 50
oxidizer, 67

Partlow, Gary and Kathy, 89-97 *passim*
palm tree shell, 113, *113*
peony shell, 113, *113*
People's Republic of China, The, 19, 65
performance standards, 15
pinwheel, 25, 41
Plimpton, George, 43, 104, 105
Polo, Marco, 23
prism: use in laser shows, 84
pyrotechnics, 9, 23, 103
pyrotechnists, 9
pyro, 9
pur, 9

Quebec, 41
Queen Victoria, *30*, 39

red-tipped comet shell, *115*
Ricci, Matteo, 23
rockets, 15, 27, 41, 109, *114*
Roman Candles, 55, 109

Rose Bowl, 49
Ruggieri family, 35, 37; American Bicentennial, 38; Claude-Fortuné, 38; in France, 36; Gaetano, 36; Green Park Show, 36; Bull of Fire, 38; shows, 38-39

safety: at a show, 89, 90, 91, 96; guidelines, 106-107; in a manufacturing plant, 64, 65, 68, 70, 72-73; organizations, 12, 15, 16, 18, 106, 107; precautions, 12, *13*, 14, 16, 18, 62, *100;* regulations, 15, 18
saltpeter, 21, 52. *See also* gunpowder
salute, 16
salute container, 72, 73
Saturn Missile, 14
set pieces, 25, 28-29, 31, 36, 39, 44, 50, 76, 77; fabrication of, 80-82, *82;* fire pictures, 30; preparation for a show, 92-93, *92, 93*, 102; as stage scenery, 102
shockwaves, 60
shooter, 86, *96*
shows, 27; aerial display, 31; choosing a site for, 78-79, *80, 95;* cleaning up after, 97, 98; cost of, 77; designing of, 79-80, *81;* finales, *97, 115;* firing methods, 78; musical programs, 78, 83; planning, 77; preparations for, 77-83, 90, *91;* sponsors of, 25, 31, 32
silver salute, 16, 110
smoke balls, 108
snake, 108
Sorgio family, 46, 48; James, 46, *47;* John, 46; Vincenzo (Jimmy the Bomb), 46
sound charge. *See* charges
Southern International Fireworks, 87, 89
spaghet. *See* fuses
sparklers, 108
spectator's guide to fireworks, 112
spider web shell, *114*
spiking, 71
Spirograph, 84
squibs, 10
stars, 58, 66; ingredients, 67; manufacturing of, *67*, 68-70, *69*
static electricity, 65
strobe shell, 45
sulfer, 21, 52. *See also* gunpowder

tetanus, 11
Thunder Buzzards, 14
Time Bandits, 103
time count track, 83
T.N.T., 15

Walt Disney World, 32, *32*
Washington, D.C., 14, 38, 45
Washington, George, 10

Washington Monument, *8*
weather, 66, 74, 89, 92, 93; for laser shows, 85
weeping willow shell, *112*
West Hampton, Long Island, 104
whistle shell, *115*
Wild Men. *See* Green Men
Wizard of Oz, The, 103
World's Fair, 1982 Knoxville, Tennessee, 44, 84

Zambelli factory, 66, 74; star mixing room, 67; cutting room, 67; spiking room, 68; flash powder shed, 72; pasting room, 74; finishing room, 74
Zambelli family, 48, 104; accidents, 65; Antonio, 48; George, 48, 50, 58, 85; George Jr., 50; Joseph, 48, 70, 72, 74; Louis, 48, 74, *75*; shows, 50, *49*, 50, 72
Zambelli Internationale, 48, *49*, 65, 66, 78, 84, *85*, 86

PICTURE CREDITS:
American Fireworks Company, 47; Brock's Fireworks, Ltd., 26, 30, 40, 41, 53, 61; Eleutherian Mills Historical Library, 28, 54; Art Gentile, K.P.C. Photography, 88, 91, 95, 96; Donal F. Holway, Photographer, 45; Library of Congress, 11; Ogatsu Fireworks Company, Ltd., 42, 112 top; New York Public Library Picture Collection, 12, 13, 22, 100; Ruggieri Inc., 20, 35, 37, *Smithsonian*, Nancy Seaman, 14, 15, 17; Southern International, 113 bottom, 115 bottom left; Tri-State Fireworks Company, 80, 81, 108, 109, 114 bottom, 115 bottom right; © Walt Disney Productions, 32, 113 top, 114 top; Zambelli Internationale, 8, 49, 56, 62, 64, 67, 69, 71, 73, 75, 79, 82, 85, 92, 93, 99, 112 bottom, 115 top.